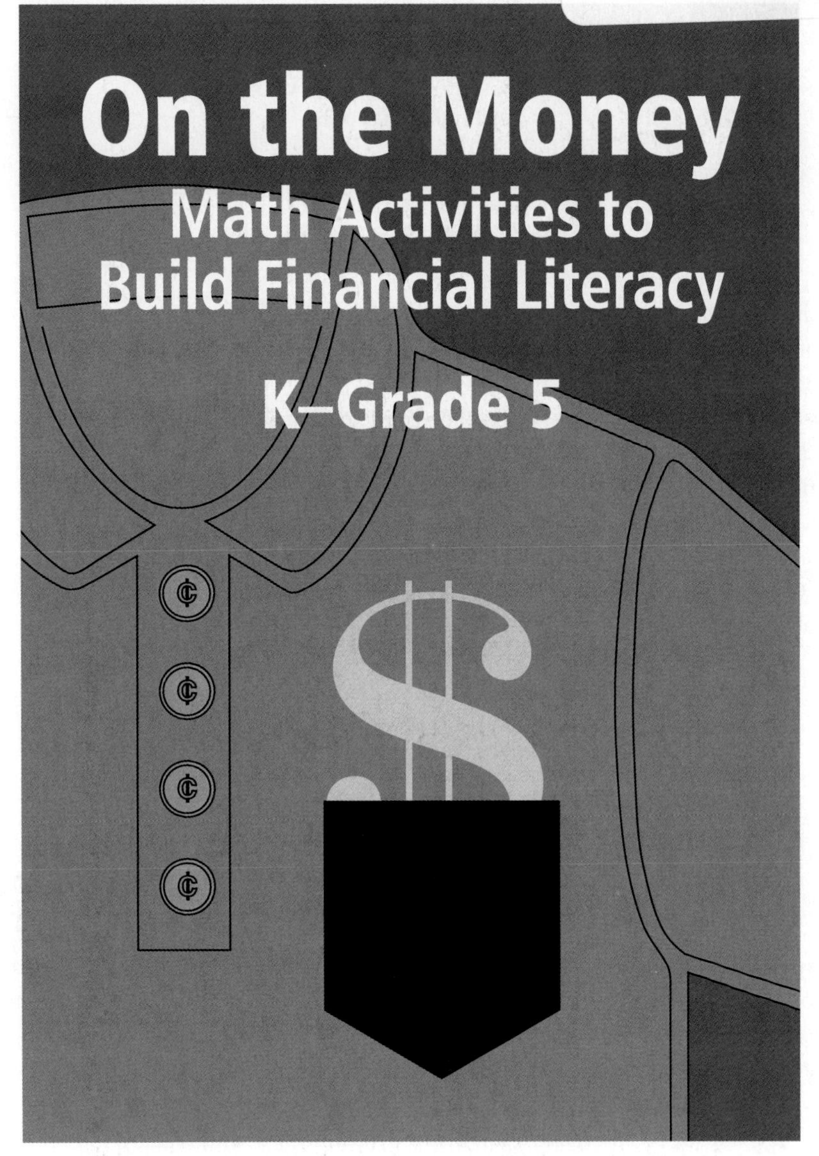

Lindsay A. Gold
University of Dayton
Dayton, Ohio

Jennifer M. Bay-Williams
University of Louisville
Louisville, Kentucky

Michael S. Houston
Riverside High School
Ellwood City, Pennsylvania

John M. Ashurst
Harlan High School
Harlan, Kentucky

Library of Congress Cataloging-in-Publication Data

Names: Gold, Lindsay A., author.
Title: On the money : math activities to build financial literacy, K–grade 5
/ Lindsay A. Gold [and three others].
Description: Reston, VA : National Council of Teachers of Mathematics, Inc., [2018].
Identifiers: LCCN 2018016047 (print) | LCCN 2018023015 (ebook) | ISBN
9781680540208 (ebook) | ISBN 9780873539968
Subjects: LCSH: Financial literacy--Study and teaching (Elementary)--Activity
programs. | Mathematics--Study and teaching (Elementary)--Activity programs.
Classification: LCC HG179 (ebook) | LCC HG179 .G644 2018 (print) | DDC 372.8--dc23
LC record available at https://lccn.loc.gov/2018016047

The National Council of Teachers of Mathematics supports and advocates
for high-quality mathematics teaching and learning for each and every student.

Printed in the United States of America

Contents

Foreword

More and more, we can see the importance of teaching our kids to be financially responsible and giving them the skills they will need to live in the real world. *On the Money: Math Activities to Build Financial Literacy, K–Grade 5* addresses this issue head-on in an approachable, fun way to really engage our children in relevant learning.

What is the big deal? The big deal is that youth (and adults) in the United States are financially illiterate. Financial literacy requires knowing how money can be spent and earned—and being able to make sound financial decisions based on this knowledge. Financial illiteracy is well documented. Just read these three sobering findings:

1. The FINRA Foundation (2013) surveyed more than 25,000 Americans using five basic financial literacy questions. The national average of this Financial Capability Study was 2.88 correct answers. (You can go to http://usfinancialcapability.org/quiz.php and take the quiz yourself to see how you measure up to others in your state and to the national average.)

2. Inceptia (2013) conducted a National Financial Capability study of 962 first-year college students, using a fifty-item test. Two-thirds scored a D or F (scoring below a 35), four in ten students did not know the meaning of "Net Pay," and more than half did not understand their credit score. And, within this same population, 60 percent reported that they did not create a budget for themselves, and 70 percent are borrowing money for college.

3. In 2014, the National Financial Educators Council (NFEC) conducted a financial literacy test on 2459 youth, ages 15–18, across forty U.S. states (NFEC 2014). Only 11 percent of students scored 80 percent or above, and the average score was 59.6 percent. This assessment has since been administered to 537 students ages 10–14, with an average score of 54 percent, and to 4916 students ages 15–18, with an average score of 60 percent. (You can log in and take this quiz, too.)

Not that many years ago, the United States came out of the largest economic meltdown since the Great Depression. Yet Americans have not learned the money lessons they should have and are still carrying heavy debt. Baby boomers and millennials alike have not gotten the emergency-warning money message. In 2015, the average U.S. household carrying debt owed more than $15,000 in credit card debt, the average mortgage debt was over $156,000, and the average student loan debt was about $33,000 (Chen 2015). More than a third (34 percent) of Americans report paying only the minimum credit card payment (FINRA Foundation 2013). Americans are also not prepared for life's money surprises. Most Americans do not have rainy-day savings to cover three months of unanticipated financial emergencies (FINRA Foundation 2013; Practical Money Skills for Life 2012). And American women, while mostly confident in managing their day-to-day budgeting (82 percent), are not confident when it comes to longer-term financial planning, such as retirement (only 37 percent are confident) and making correct investments (28 percent confident) (Fidelity Investments 2015). This just scrapes the surface of the many reports that confirm the lack of financial literacy in the United States.

How do our U.S. youth stack up in the global market? According to Visa's Global Financial Literacy Barometer (Practical Money Skills for Life 2012), the United States ranked fourth of the twenty-eight participating countries in overall financial literacy. (Brazil topped the list as having the most financially savvy folks, followed by Mexico and Australia—perhaps a subtle hint that *On the Money* should be translated!) And yet, when asked, "To what extent would you say that teenagers and young adults in your country are adequately prepared to manage their own money?" the United States was ranked twenty-seventh (second to last). Across these countries, and in the United States, the average age that respondents thought financial literacy education should begin was about eleven—middle school age! What happens when our children do not get this knowledge? The current financial illiteracy is passed on to adulthood.

In the mid-1980s, when I was president of the First Women's Bank, I felt that a financial education train wreck was heading our way. I saw women disempowered from handling their own money matters, and my research showed it was because we were never taught anything about money as kids. I went to look for books to teach my own kids. There weren't any . . . so I wrote one book, and then another one, and I eventually worked to create the topic of Kids & Money as something we all needed to address. I advocated for banks, schools, and families to listen, and fortunately there are now many great thought leaders who are advocating for the financial literacy of American youth.

What is happening today? We all know that financial literacy is crucial to the welfare of our families, our communities, our nation, and even our world. The final report of the President's Advisory Council on Financial Capability (2013) states that the four major recommendations—

> share a common element: they acknowledge that financial capability is not a "stand-alone" topic to be isolated from the rest of our lives. Financial capability must be woven into the fabric of our lives—into our homes, our schools, our workplaces, our communities, even the design and regulation of the financial products and services we use." (p. III)

I love that their first recommendation is to focus on our youth and to integrate personal finance into the teaching of the Common Core State Standards for English and mathematics. That is why *On the Money* is so crucial to our children and should be required learning in every school system in our nation . . . NOW. Its authors have made the connections between math and real life. Let's face it—now more than ever our kids need these skills in order to live happy and productive lives. The ideas in this book really make earning, saving, and spending relevant. *On the Money* connects students to finding sale discounts, devising saving strategies, and other real-life skills they will need to use every day. The book even deals with what I call "Paycheck Shock." I remember getting my first paycheck and looking at it in shock, not understanding the difference between "net" and "gross." In dismay, I quipped, "Who is FICA and how dare he take money out of my paycheck!" With the amazing activities you'll find here, math education comes alive, with students actually figuring out how a real paycheck works and how to calculate the relevant percentages. My only negative comment as it relates to this book is, "Where were you when my kids were growing up?"

On the Money turns our students' world into their classroom. I thank its authors and all the teachers and educators who will use this book for their work on behalf of our children.

Neale Godfrey is the best-selling author of a number of books on financial literacy, including *Money Doesn't Grow on Trees: A Parent's Guide to Raising Financially Responsible Children; Money Still Doesn't Grow on Trees; The Ultimate Kids' Money Book;* and *A Penny Saved: Teaching Your Children the Values and Life Skills They Will Need to Live in the Real World.*

References

Chen, Tim. "American Household Credit Card Debt Statistics: 2015" (2015). http://www.nerdwallet.com/blog/credit-card-data/average-credit-card-debt-household/.

Fidelity Investments. "Money Fit Women Study: Executive Summary" (2015). https://www.fidelity.com/bin-public/060_www_fidelity_com/documents/women-fit-money-study.pdf.

FINRA Investor Education Foundation. *National Financial Capability Study* (2013). http://www.usfinancialcapability.org.

Inceptia. "First-Year College Students Score Poorly in Basic Financial Literacy, Inceptia Survey Reports" (2013). https://www.inceptia.org/resource-center/news/jan-22-2013/.

National Financial Educators Council (NFEC). "National Financial Literacy Test Results" (2014). http://www.financialeducatorscouncil.org/financial-literacy-research/.

Practical Money Skills for Life. "Federal Reserve Bank of Chicago and Visa Inc. Host Sixth Annual Global Summit on Financial Literacy" (2012). http://www.practicalmoneyskills.com/about/press/releases_2012/0423.php.

President's Advisory Council. "President's Advisory Council on Financial Capability: Final Report" (2013). http://www.treasury.gov/resource-center/financial-education/Documents/PACFC final report revised 22513 %288%29_R.pdf.

Introduction

When a person hears "literacy" related to the elementary grades, what usually comes to mind is reading. While reading is a core component of the primary curriculum, it cannot be the only focus for literacy. *On the Money: Math Activities to Build Financial Literacy, K–Grade 5* addresses the college- and career-readiness concepts and skills of financial literacy as meaningfully connected to mathematical proficiencies and content for elementary-age students.

Becoming financially literate is not an "absolute state; but rather, a continuum of abilities that is subject to variables throughout the life cycle. It is an evolving state of competency that enables individuals to respond effectively to ever-changing personal and economic circumstances" (Jump$tart Coalition for Personal Financial Literacy 2017, p. 1). The goal of the *National Standards in K–12 Personal Finance Education* (from here on referred to as the Jump$tart Standards) is to develop students' ability to find answers and seek solutions for making sound decisions regarding finances.

Connecting to financial literacy is both necessary and manageable. Combining the Jump$tart Standards along with the Common Core State Standards for Mathematics (CCSSM; National Governors Association Center for Best Practices and Council of Chief State School Officers [NGA Center and CCSSO] 2010), *On the Money* provides engaging and challenging lessons that develop financial literacy appropriate for young children while also developing grade-level mathematics. These lessons can be embedded in your already established curriculum at the time when you are teaching the relevant mathematics, or they can be used as enrichment or extension in addition to your regular curriculum.

The chapters are formatted in a way that allows teachers to make adjustments to levels and difficulty based on their individual students' needs. Chapters include the following sections:

- **The Bottom Line:** This is a brief summary of the chapter's contents and takeaways.
- **Standards Met in This Chapter:** This section lists the specific CCSSM and Jump$tart standards related to the chapter's lesson. The CCSSM standards include the Standards for Mathematical Practice, while the Jump$tart standards include a breakdown of the individual standards and knowledge statements related to the financial literacy content.
- **Future Value:** This section is a look into how the chapter content will affect your students in their future educational and financial experiences. It is divided into subsections on **Mathematics, Financial Literacy,** and **Student Knowledge and Experience** that describe the mathematical and financial literacy concepts that the chapter will encompass, as well as the knowledge and experiences that elementary students may bring to the lesson and take away from it.

Chapters also include a **Lesson Plan** with various parts that are meant to be accomplished on the teacher's timeline, not necessarily in one class setting. Each lesson includes learning targets, resources and tools, preparation tips, suggestions of related children's literature, and key financial and mathematical terms. Lessons also include these sections:

- **Money Talks:** This opening section provides a way to talk with your students about financial concepts and skills and how these play a role in their everyday lives.

- **Sales Pitch:** This is the lesson opener, usually including a reference to a children's literature selection.

- **The Fine Print:** Here you will find the parts of the lesson. These parts are meant to be covered as the teacher sees fit and can be adapted for various grade levels and abilities. Occasional notes in the margins include interesting "Did you know?" facts and links to the Standards for Mathematical Practice.

- **Closing the Deal:** This is the wrap-up for the lesson. Take time to revisit the learning targets and the vocabulary associated with the chapter content.

- **Getting Your Money's Worth:** As written, not every lesson will meet the needs of every child, so this section provides suggestions on how lessons can be modified or varied.

- **References and Resources:** Here you will find citations and sources for materials used within the chapter. In the back of the book, you will also find an extensive resource list for children's literature based on grade level or topic, as well as helpful technology links.

At the end of each chapter, you will find **Activity Sheets** and **Resource** pages. The activity sheets are designed to be completed by students as part of the lesson, while the resource pages include images, cards, and other materials that can be used during the lesson. All of these pages are available as downloadable PDFs on NCTM's More4U website (http://www.nctm.org/more4u) by using the code on the title page of this book.

Why Teach Financial Literacy in K–Grade 5?

> Children develop an interest in money and begin to learn financial basics well before entering school. (Jump$tart Coalition 2017)

Making decisions about how to use money, how to make money, and when and how you want to save money are relevant topics for elementary-age students. Some argue that the elementary curriculum is too full to make room for "new" ideas or additional topics. But financial literacy is not a completely separate and distinct topic; it is a subject with inherent connections to mathematics and social studies. Many organizations and leaders advocate for addressing financial literacy in the early elementary years (Birbili and Kontopoulou 2015). For example:

> Children learn about money from many sources. Long before they enter school, they observe adults using money and buying things. They watch television daily and see thousands of commercials each year. Like it or not, money is a part of your preschooler's life. (Credit Union National Association, para. 5)

That money is a relevant, interesting topic for young children is one argument for why financial literacy should be included in K–5 education. From a mathematics teaching perspective, it is important that students see mathematics as useful and meaningful. Financial literacy topics can provide an excellent opportunity to connect to students' interests and prior knowledge and to help them see how mathematics can be used to navigate important life decisions.

Another compelling reason to teach financial literacy in our schools right from the start is that there are so many adults who are not financially literate, and this can lead to significant challenges and losses. Too many people do not understand how credit works, do not grasp the long-term implications of short-term loans (such as payday loans and credit cards), and are not planning for the future or for big purchases. Becoming financially knowledgeable and responsible is absolutely necessary to personal well-being. It is this dire need that has led many financial organizations to advocate for the inclusion of financial literacy starting with very young students. Beyond the Jump$tart Standards, which include specific knowledge statements for kindergartners, the following statements describe the need for early financial literacy instruction:

- The President's Advisory Council on Financial Capability for Young Americans (2015) stated, "While evidence is scarcer on the financial capability of children under age 15, we believe that today's teens and young adults who lack financial capability not long ago were children who did not have opportunities to gain financial knowledge and skills. . . . Despite the clear need for financial education for young people to make sound decisions and have more positive outcomes, too many children do not obtain financial education early in their lives" (p. 8).
- The National Association of State Boards of Education (NASBE) stated, "The earlier a student begins learning these [financial literacy] concepts, the more opportunities schools will have to impact behavior" (2006, p. 20).
- The Assets and Education Initiative (2013) summarizes research suggesting that "financial resources have their strongest effect on children's educational outcomes early on in the child's life, not at the point of college entry" (p. 16).

How Do You Teach Financial Literacy in Mathematics?

The authors of this book believe in an integrated curriculum approach to teaching, one in which financial literacy can absolutely be a component of the mathematics curriculum. As demonstrated in the book's chapters, financial literacy concepts and skills can be interwoven into most domains of the mathematics curriculum. In the *On the Money* series, we refer to such lessons as Financial Literacy Mathematics (FLM) Lessons. An FLM lesson does more than simply teach facts about coins, use pennies and dimes for place value or counting, or include story problems with money as a context. An FLM lesson has students learn grade-level mathematics while they also learn age-appropriate financial literacy. Conversely, a financial literacy lesson is not a mathematics lesson unless it includes goals and outcomes related to mathematics. Consider an FLM lesson as a lesson that addresses content in the overlapping region of a Venn diagram:

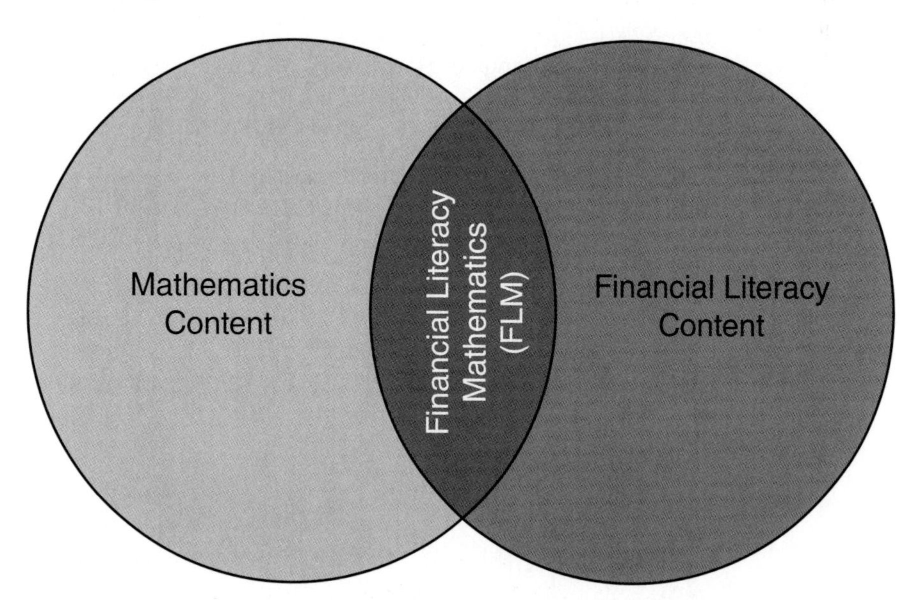

As this visual implies, not all mathematics is directly related to financial literacy (e.g., angle measures in geometric shapes), and many aspects of financial literacy are not directly related to mathematics (e.g., explain why something borrowed must be returned [kindergarten] or identify situations when people might pay for something over time [grade 4]). But there is a significant amount of overlap between the two domains. The content for the lessons in this book lie within that intersection of mathematical and financial literacy knowledge.

Before the chapters begin, the following three pages of lists and tables give a detailed breakdown of each chapter and its alignment to both the Common Core State Standards for Mathematics and the National Standards in K–12 Personal Finance Education, 4th Edition, 2nd Printing (2017).

Acknowledgments

We would like to thank the many people who have helped to prepare this book and its supplementary materials. Their encouragement, constructive criticisms, and suggestions have been invaluable to us. A special thanks to Cathryn Rodgers and Becky Byer for their consultation on content and technology files, and to our families for their love, support, and patience throughout this journey.

References

Assets and Education Initiative. "Building Expectations, Delivering Results: Asset-Based Financial Aid and the Future of Higher Education." In *Biannual Report on the Assets and Education Field*, edited by William Elliot. Lawrence, Kans.: Asset and Education Initiative, 2013.

Birbili, Maria, and Melanthia Kontopoulou. "Financial Education for Preschoolers: Preparing Young Children for the 21st Century." *Childhood Education* 91 (2015): 46–53.

Credit Union National Association (2015). *Thrive by Five* [web page]. http://www.cuna.org/thrivebyfive/.

Jump$tart Coalition for Personal Financial Literacy. *National Standards in K–12 Personal Finance Education*. Fourth Edition, Second Printing. Washington, D.C.: Jump$tart Coalition for Personal Financial Literacy, 2017.

National Association of State Boards of Education. *Who Will Own Our Children? The Report of the NASBE Commission on Financial and Investor Literacy (2006). http://www.nasbe.org/wp-content/uploads/SG_Financial_Literacy_*2006.pdf.

National Governors Association Center for Best Practices and Council of Chief State School Officers (NGA Center and CCSSO). *Common Core State Standards for Mathematics.* Washington, D.C.: NGA Center and CCSSO, 2010.

President's Advisory Council on Financial Capability for Young Americans. Final Report, June 2015 (2015). Washington D.C.: President's Advisory Council on Financial Capability for Young Americans. https://www.treasury.gov/resource-center/financialeducation/Documents/PACFCYA%20Final%20Report%20June%202015.pdf.

CCSS Standards for Mathematical Practice

(National Governors Association Center for Best Practices and Council of Chief State School Officers 2010)

MP.1 Make sense of problems and persevere in solving them.
MP.2 Reason abstractly and quantitatively.
MP.3 Construct viable arguments and critique the reasoning of others.
MP.4 Model with mathematics.
MP.5 Use appropriate tools strategically.
MP.6 Attend to precision.
MP.7 Look for and make use of structure.
MP.8 Look for and express regularity in repeated reasoning.

Aligning Chapters to the Common Core State Standards for Mathematics (CCSSM)

CCSSM Domains and Mathematical Practices						
Chapter	**Counting & Cardinality**	**Operations & Algebraic Thinking**	**Numbers & Operations in Base 10**	**Number & Operations–Fractions**	**Measurement & Data**	**Mathematical Practices**
1	K.CC.B.4				K.MD.A.1 K.MD.B.3 1.MD.C.4	3, 6
2	K.CC.A.3 K.CC.B.4 K.CC.B.5 K.CC.C.6					2
3		K.OA.A.1 1.OA.B.4	2.NBT.A.2			3, 4
4	K.CC.A.1	K.OA.A.2 1.OA.A.1 1.OA.B.3 2.OA.A.1	1.NBT.C.4 2.NBT.A.2 2.NBT.B.7 3.NBT.A.2		2.MD.C.8	4, 7
5		3.OA.C.7 3.OA.D.8 4.OA.A.3				1, 5, 6

Chapter	Counting & Cardinality	Operations & Algebraic Thinking	Numbers & Operations in Base 10	Number & Operations–Fractions	Measurement & Data	Mathematical Practices
6		3.OA.D.8 3.OA.D.9 4.OA.A.3 4.OA.C.5 5.OA.B.3				1, 2, 4
7			1.NBT.B.2 1.NBT.C.5 2.NBT.A.1 2.NBT.B.5			5, 7
8			1.NBT.B.2.c 1.NBT.C.5 1.NBT.C.6 2.NBT.B.8			7, 8
9			3.NBT.A.2 4.NBT.A.1 4.NBT.B.4 5.NBT.A.3 5.NBT.B.7			4, 5, 7
10			4.NBT.B.4 5.NBT.B.7	4.NF.C.6 5.NF.B.6		4, 6
11		4.OA.C.5 5.OA.B.3		4.NF.C.6		4, 7, 8
12				3.NF.A.1 4.NF.B.3.a 5.NF.B.3		2, 4, 7
13			3.NBT.A.2 4.NBT.B.4 5.NBT.A.3 5.NBT.B.7	4.NF.C.6		6
14					1.MD.C.4 2.MD.D.10 3.MD.B.3	1, 6
15					K.MD.B.3 1.MD.C.4	3, 5
16					1 K.MD.A.1 1.MD.A.2 2.MD.A.2	4, 5, 6
17					3.MD.B.3 4.MD.B.4 5.MD.B.2	3, 4

National Standards in K–12 Personal Finance Education, 4th Edition, 2nd Printing

(Jump$tart Coalition for Personal Financial Literacy 2017)

Spending and Saving

Overall Competency

Apply strategies to monitor income and expenses, plan for spending and save for future goals.

Standard 1. Develop a plan for spending and saving.

Standard 2. Develop a system for keeping and using financial records.

Standard 3. Describe how to use different payment methods.

Standard 4. Apply consumer skills to spending and saving decisions.

Credit and Debt

Overall Competency

Develop strategies to control and manage credit and debt.

Standard 1. Analyze the costs and benefits of various types of credit.

Standard 2. Summarize a borrower's rights and responsibilities related to credit reports.

Standard 3. Apply strategies to avoid or correct debt management problems.

Standard 4. Summarize major consumer credit laws.

Employment and Income

Overall Competency

Use a career plan to develop personal income potential.

Standard 1. Explore job and career options.

Standard 2. Compare sources of personal income and compensation.

Standard 3. Analyze factors that affect net income.

Investing

Overall Competency

Implement a diversified investment strategy that is compatible with personal financial goals.

Standard 1. Explain how investing may build wealth and help meet financial goals.

Standard 2. Evaluate investment alternatives.

Standard 3. Demonstrate how to buy and sell investments.

Standard 4. Investigate how agencies protect investors and regulate financial markets and products.

Risk Management and Insurance

Overall Competency

Apply appropriate and cost-effective risk management strategies.

Standard 1. Identify common types of risks and basic risk management methods.

Standard 2. Justify reasons to use property and liability insurance.

Standard 3. Justify reasons to use health, disability, long-term care and life insurance.

Financial Decision Making

Overall Competency

Apply reliable information and systematic decision making to personal financial decisions.

Standard 1. Recognize the responsibilities associated with personal financial decisions.
Standard 2. Use reliable resources when making financial decisions.
Standard 3. Summarize major consumer protection laws.
Standard 4. Make criterion-based financial decisions by systematically considering alternatives and consequences.
Standard 5. Apply communication strategies when discussing financial issues.
Standard 6. Analyze the requirements of contractual obligations.
Standard 7. Control personal information.
Standard 8. Use a personal financial plan.

Aligning Chapters to the Jump$tart Financial Literacy Standards

	Categories and Numbered Standards of the Financial Literacy Standards					
Chapter	Spending and Saving	Credit and Debt	Employment and Income	Investing	Risk Management and Insurance	Financial Decision Making
1	3					
2	1, 3					
3					KS	
4	3					4
5	3		2			
6	KS					1
7	1, 2					
8		3				
9			2			
10	KS		KS			
11				1		
12	1					
13	3					
14			1, 2			
15						7
16	3					
17			1, 2			

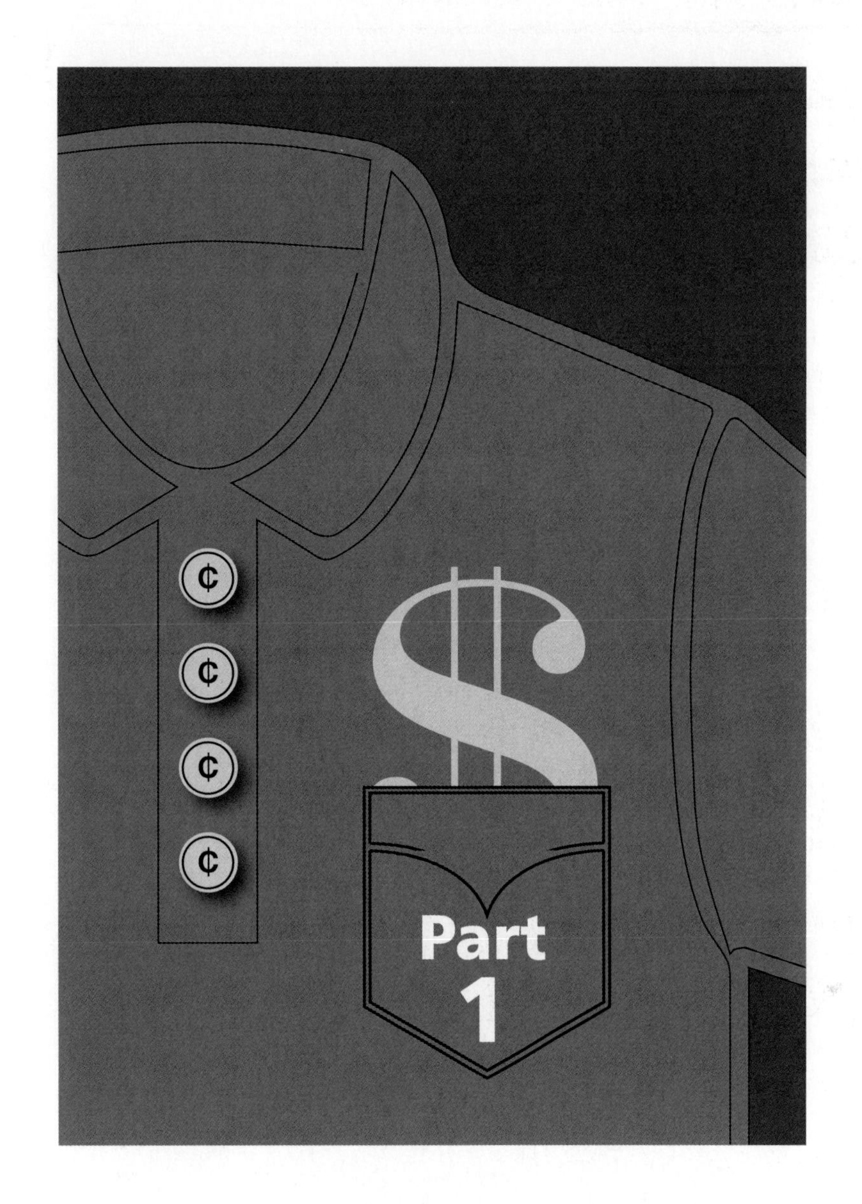

Counting and Cardinality

Chapter

Coin Attributes

The Bottom Line

Coins have various attributes that cause them to be distinctly different from one another.

Students sort and classify coins, count the number of coins in each category, and describe and compare coins based on their measurable attributes.

Standards Met in This Chapter

CCSSM Standards and Practices

K.CC.B.4. Understand the relationship between numbers and quantities; connect counting to cardinality.

K.MD.A.1. Describe measurable attributes of objects such as length or weight. Describe several measurable attributes of a single object.

K.MD.B.3. Classify objects into given categories; count the numbers of objects in each category and sort the categories by count.

1.MD.C.4. Organize, represent, and interpret data with up to three catagories; and and answer questions about the total number of data points, how many in each category, and how many more or less are in one category than in another

MP.3: Construct viable arguments and critique the reasoning of others.

MP.6: Attend to precision.

(National Governors Association Center for Best Practices and Council of Chief State School Officers [NGA Center and CCSSO] 2010)

Jump$tart Standards

Spending and Saving: Apply strategies to monitor income and expenses, plan for spending and save for future goals.

Standard 3. Describe how to use different payment methods.

Knowledge Statements:

Overall: Paper money and coins have different values

Standard 3: Sort coins and paper money by appearance and name.

(Jump$tart Coalition for Personal Financial Literacy 2017)

Future Value

The most recognized coin in the American monetary system is the penny. The copper color of a penny makes it distinguishable from other U.S. coins; it is also the least valuable coin, at only one cent. One of the first experiences that children will have with money is noticing these differences in attributes, which make each coin unique.

In this chapter, students will explore the measurable attributes of the penny, nickel, dime, and quarter and will classify them based on these similarities and differences. After the following activities, students should be able to distinguish these specific categories and describe them based on how the coins are sorted.

The following sections describe the mathematical and financial literacy concepts that this lesson will encompass, as well as the knowledge and experiences that elementary students may bring to the lesson and take from it on its completion.

Did you know? Before the 18th century was over, the U.S. Mint began adding ridges to the coins' edges, a process called "reeding," in order to make it impossible to shave them down without the result being obvious. As a side benefit, the reeded edges also made coin design more intricate and counterfeiting more difficult (Maranzani 2012).

Mathematics

Like geometric shapes, coins have characteristics and attributes that make them distinct from one another. These characteristics include color, face picture, value, and the feel of the outer edge of the coin. Other measurable attributes (K.MD.A.1) include diameter, thickness, and weight. Students can describe these attributes with mathematical vocabulary such as larger, smaller, thicker, thinner, heavier, and lighter. Through the activities in this chapter, students will use reasoning to compare various coins and determine their defining characteristics. They will also describe measurable attributes of each coin and compare them to one another (K.CC.B.4). When comparing, students must attend to precision (MP.6), specifically in measurable attributes.

Students' ability to classify objects is fundamental for future data collection skills (K.MD.B.3). By providing students with multiple opportunities to classify and categorize sets, students can discover that sometimes there are multiple ways to classify the same collection of coins. Students will increase their mathematical vocabulary through descriptive words (MP.3), be able to organize data, and answer questions based on their findings (1.MD.C.4).

Once students have determined data sets, counting and cardinality standards (K.CC.B.4) are reinforced through one-to-one correspondence between the numeral representation and the number of coins in the set.

Financial Literacy

A large part of improving financial literacy at a young age is recognizing that different coins have different values. Students must be able to identify each coin and learn to distinguish how much each is worth. Through these repetitive experiences, students begin to identify the penny, nickel, dime, and quarter and label them with their appropriate value using the cent (¢) sign. In

Did you know?
The *cent*, frequently represented by the cent sign (a miniature letter "c" crossed by a vertical or diagonal stroke: ¢; or a simple "c") is a monetary unit that equals 1/100 of a dollar or other decimal currency unit. The word "cent" derives from the Latin word *centum*, meaning hundred (dictionary.com n.d.).

doing so, students start to realize that some coins are worth more than others, and this realization can be a precursor to exchanging less valued coins for their larger counterpart—e.g., five pennies for one nickel.

Student Knowledge and Experience

This lesson builds on students' skills of sorting and counting objects and emerging understanding of data collection and interpretation. Some students will first see the difference in color, thus separating pennies from the silver color of the other coins. Others will notice the size difference. This recognition will lead to an important discussion in that size does not have a direct correlation with value. The relationship of size to value is a common student misconception and must be addressed. This rich conversation could lead to a student-led investigation during a different lesson on why the coins were created with specific attributes.

Once students have sorted and described their sets, they can practice their counting and cardinality skills by matching or writing the number of coins with the correct numeral. Students will continue to develop their one-to-one correspondence as they match each item with its corresponding numeral and identify the cardinality for each set.

Lesson Plan

Learning Targets

- Classify coins by their defining attributes and characteristics.
- Connect counting to cardinality by identifying how many coins are in each set.
- Organize and interpret the coin data to ask and answer questions.

Resources and Tools

- **Resource 1.1a and 1.1b:** Money Jars (one sheet of each for each student or pair of students)
- **Resource 1.2:** Number Cards 1–20
- Coins to sort (pennies, nickels, and dimes; quarters depending on appropriateness)
- Money jar (yogurt containers, baby food jars, and applesauce cups all work well for this activity)

Preparation Prior to Lesson

Part 1: Prepare enough copies of Resource 1.1a and 1.1b as needed for each student or pair of students. Laminate a class set for reuse. Place a collection of coins into each money jar depending on student appropriateness.

Part 2: Make copies of Resource 1.2 for students who are not yet able to write the corresponding numeral.

Children's Literature Option

The Coin Counting Book by Rozanne Williams (2001). For other possible resources, see the Literature for Financial Literacy list on p. 203.

Key Language

attribute: A characteristic of an object.

sort: To group based on an attribute of an object.

Money Talks

Ask children if they, or their parents, have a "money jar." Show them a small jar full of coins. Ask them what a money jar might be used for or how money can "grow" in the jar. Talk with the students about how one might count the money in the jar. Lead them to the realization that it is easier to count money if you **sort** it into the different coins.

Sales Pitch

Explain to the students that coins have different attributes. This means that the coins have different characteristics unique from one another. Give each student, or pair of students, a "jar" with a predetermined amount of coins. For example, quarters would not be used in a kindergarten classroom, and the amount placed in each jar can vary based on ability. Have students carefully dump their money onto their desks. Give them time to explore the various coins, and record their noticings and wonderings on the board. As students are noticing differences in the coins, explain that these individual characteristics are called **attributes.** Ask the following questions:

- How do the coins look different?
- How do the coins feel different?
- What are some attributes you notice that are unique to the penny, nickel, and dime?
- What are some different ways we could sort the coins?

Explain that the penny is the most easily recognized coin because of its color. The nickel and dime are similar in color, but the nickel is larger, thicker, and has a smooth edge, whereas the dime is smaller, thinner, and has a ridged edge. Some students will have the background knowledge regarding the value of each coin and this can be an extension for further practice, but first focus on the visible elements. Make a list of defining attributes on the board for students to refer to later in the activity. Discuss which attributes are measurable (such as size and weight) versus those that are not (such as color and ridges).

The Fine Print

Part 1: Sorting

Pass out **Resources 1.1a and 1.1b** to the students. Explain to them that they are going to sort their coins into money jars. The students may choose whether to sort into two or three categories. (Fig. 1.1a shows two categories, while fig. 1.1b shows three.) The students will choose an attribute by which to sort and place the coins in the corresponding jars. They will then switch places with a partner to examine how the other person sorted and determined the defining attribute. After carefully inspecting the coins in the jars, have the students make notes about how the coins might be sorted.

Standards for Mathematical Practice: Students develop MP.3 when they determine how the coins are sorted and can justify their reasoning.

They could write or draw a picture to express their thinking. For instance, a student might write that the coins in a certain jar are brown, or thick. Once students have taken notes on the characteristics of the coins in each jar, have them choose one attribute that could be used to define the sorted set. Encourage students to refer back to the attribute list made earlier in the lesson.

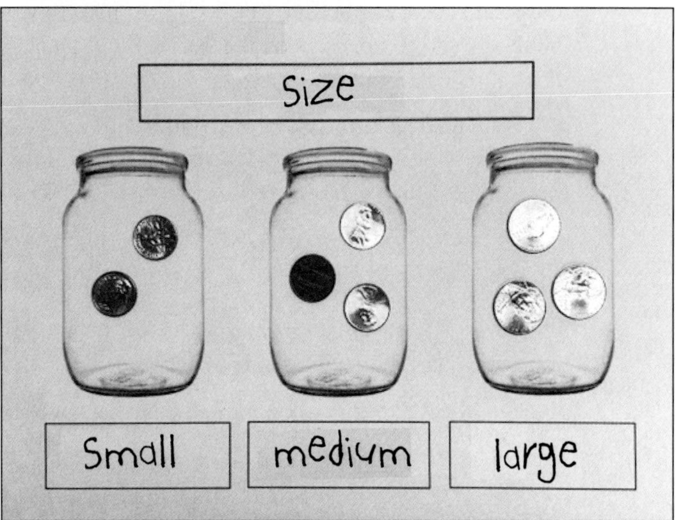

Figs. 1.1a and 1.1b. Coins sorted by two and three attributes

Part 2: Counting

After the students have had multiple opportunities to sort and classify the coins by their attributes, explain to them that they are going to practice counting and writing numerals. As students work toward counting to 100, they must recognize the 0–9 digit sequence and develop the transition signals for generating a new series of ten. They must also be able to use one-to-one correspondence skills while keeping track of which elements have been counted. Have the students begin by sliding their money jar resource sheet to the top of their workspace. Next, have them bring all the coins down from the first jar and place them in front of them. Have students arrange the coins into a line. As students are counting, have them place a finger on each individual coin and slide it up to make a new line. One of the difficulties that children contend with when counting is keeping track of which coins have been counted and which have not. By physically separating the coins, students will have an effective strategy for distinguishing between the two (fig. 1.2).

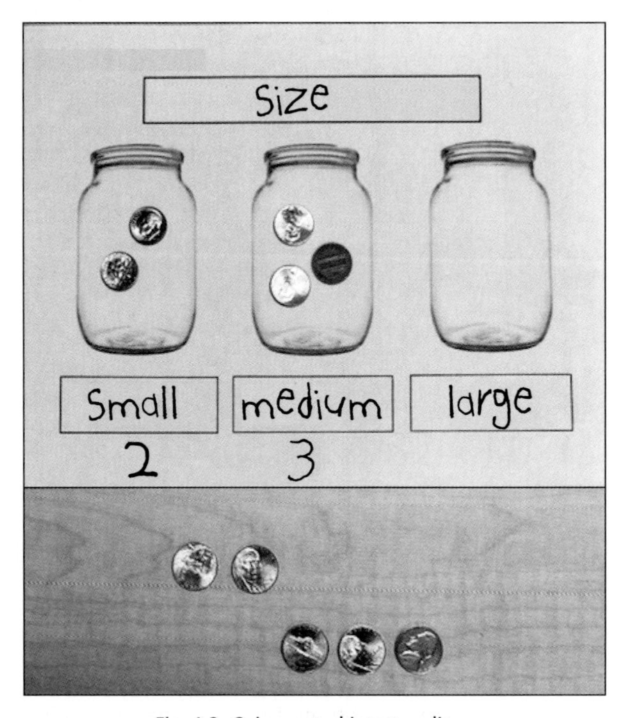

Fig. 1.2. Coins sorted into two lines

Standards for Mathematical Practice: As students count the sets of coins, they are attending to precision when recounting for accuracy (MP.6).

Students can then record the number of coins in each jar by either writing the number or matching using a number card **(Resource 1.2: Number Cards 1–20).** Encourage students to count each set more than one time to validate their answers.

Closing the Deal

Share with students that this activity is the start to knowing how much total value of money is in the money jar. Ask the following questions:

- What did you notice about the jars as you were counting the coins?
- Did some attributes tend to have more coins than others?
- Why do you think they did or did not?

Place two or three new jars in front of the students, with coins sorted by an attribute that has not been discussed yet (for example, as in fig. 1.3, where the coins are sorted by heads and tails). Ask the following questions:

- What do you notice about the contents of each jar?
- How do you think I sorted my coins?

As a class, count the number of coins in each jar. Ask the following questions:

- Which jar has the most coins?
- Which jar has the least?
- If I were trying to save my money, would I add more coins to the jar or take some away?

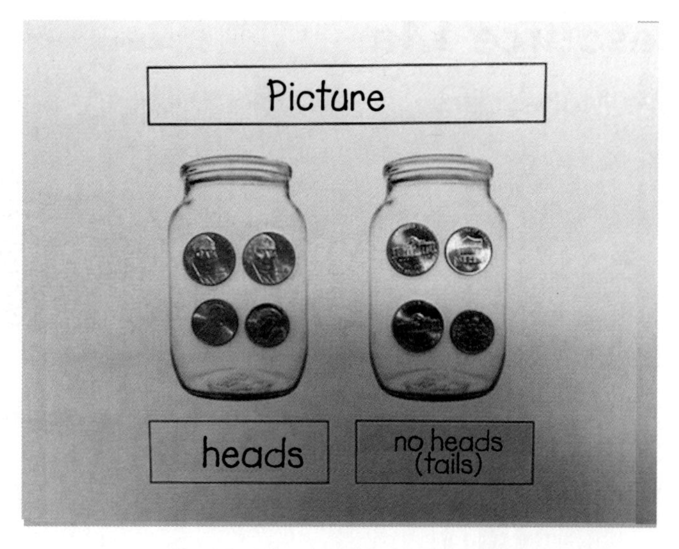

Fig. 1.3. Coins sorted by heads or tails

Close the lesson by having students share something they think is worth saving for. One possible extension is to start a class money jar, adding coins as positive reinforcement in saving and working toward a common goal.

Getting Your Money's Worth

This activity can be modified for various grades through the amount of money in each jar and adding the attribute of value. Continue to revisit this activity through a class money jar, using mathematical strategies such as repeating counting and determining how much more money is needed until a goal is reached.

References and Resources

Dictionary.com [website] (n.d.). http://www.dictionary.com/browse/cent.

Jump$tart Coalition for Personal Financial Literacy. *National Standards in K–12 Personal Finance Education.* Fourth Edition. Second Printing. Washington, D.C.: Jump$tart Coalition for Personal Financial Literacy, 2017.

Maranzani, Barbara. "8 Things You May Not Know about Money." History.com, 2012. http://www.history.com/news/history-lists/8-things-you-may-not-know-about-money.

National Governors Association Center for Best Practices and Council of Chief State School Officers (NGA Center and CCSSO). *Common Core State Standards for Mathematics.* Washington, D.C.: NGA Center and CCSSO, 2010.

Williams, Rozanne Lanczak. *The Coin Counting Book.* Watertown, Mass.: Charlesbridge, 2001.

Resource 1.1a

Two Money Jars

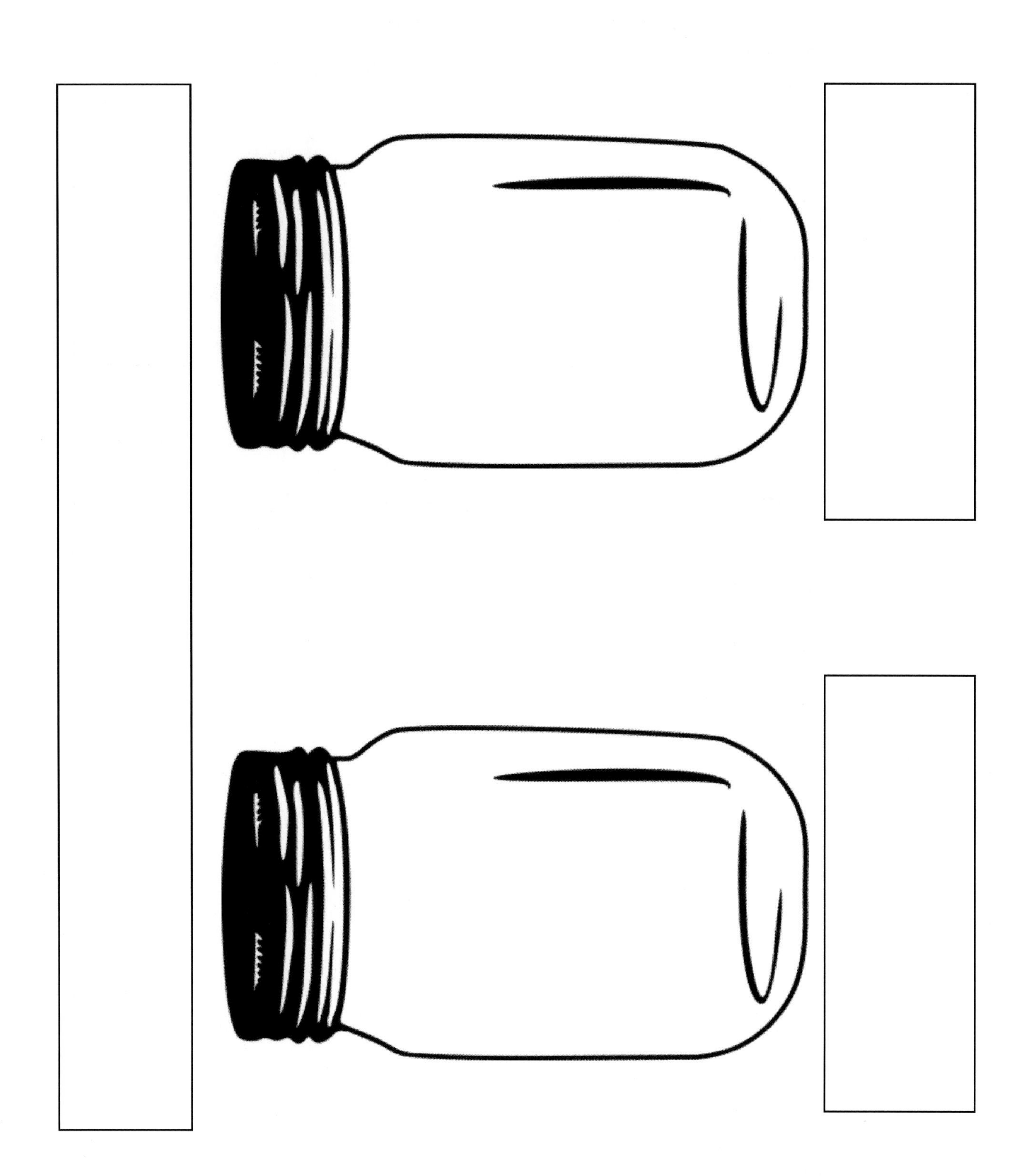

Resource 1.1b

Three Money Jars

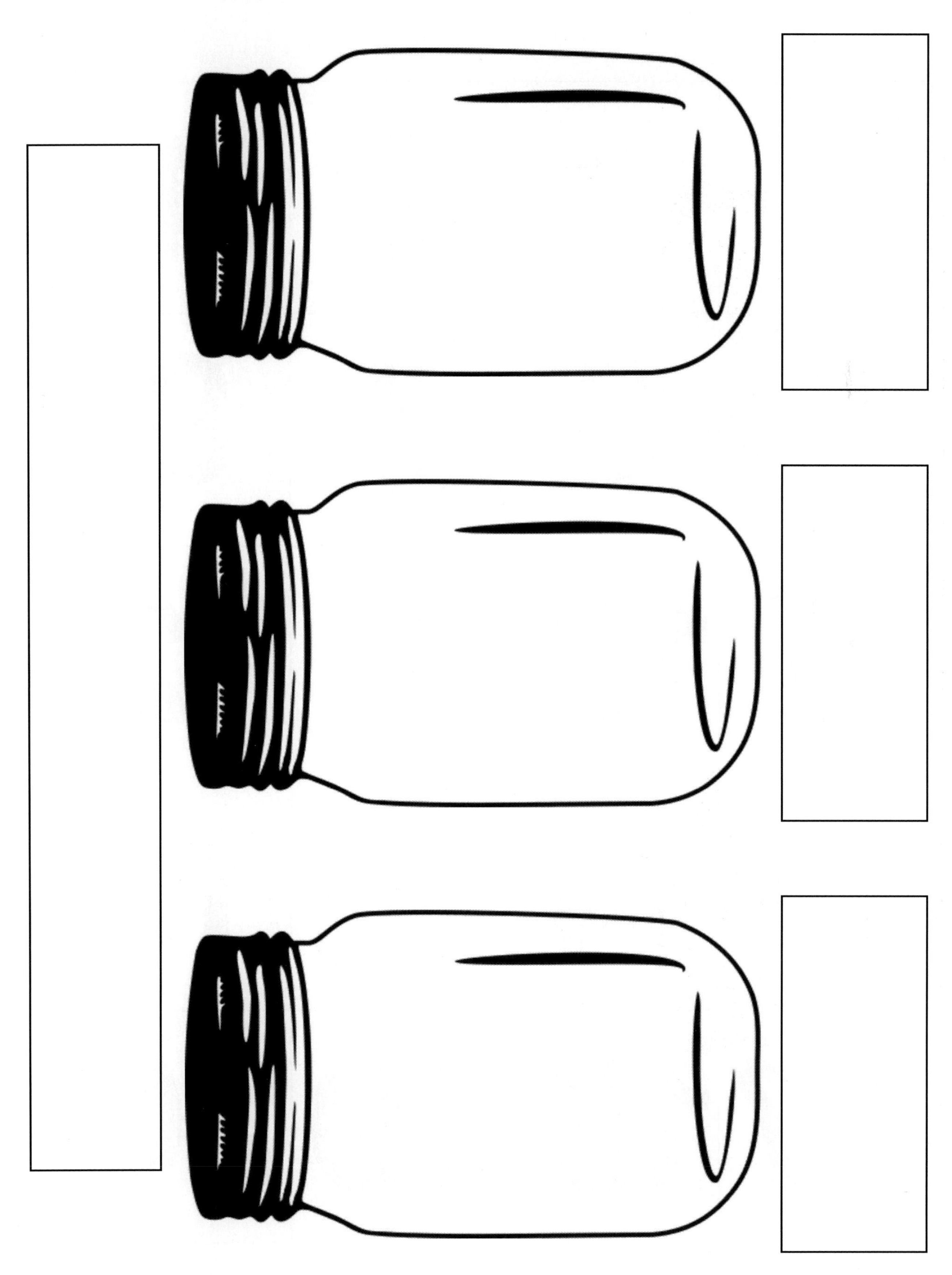

Resource 1.2

Number Cards 1–20

1	2	3	4
5	6	7	8
9	10	11	12
13	14	15	16
17	18	19	20

Chapter

A Pocketful of Pennies

The Bottom Line

A collection of coins can be counted and used to purchase goods.

Students count and compare a collection of pennies and then explore how they can be used to buy things.

Standards Met in This Chapter

CCSSM Standards and Practices

K.CC.A.3: Write numbers from 0 to 10. Represent a number of objects with a written numeral 0–20 (with 0 representing a count of no objects).

K.CC.B.4: Understand the relationship between numbers and quantities; connect counting to cardinality.

K.CC.B.5: Count to answer "how many?" questions about as many as 20 things arranged in a line, rectangular array, or a circle, or as many as 10 things in a scattered configuration: given a number from 1–20, count out that many objects.

K.CC.C.6: Identify whether the number of objects in one group is greater than, less than, or equal to the number of objects in another group, e.g., by using matching and counting strategies.

MP.2: Reason abstractly and quantitatively

(National Governors Association Center for Best Practices and Council of Chief State School Officers [NGA Center and CCSSO] 2010)

Jump$tart Standards

Spending and Saving: Apply strategies to monitor income and expenses, plan for spending and save for future goals.

Standard 1. Develop a plan for spending and saving.

Standard 3. Describe how to use different payment methods.

Knowledge Statements:

Overall:

- People trade money to buy goods or services (tasks performed by others).
- Different goods or services have different prices.

Standard 1: Decide uses for personal funds.

Standard 3: Count items up to 10.

(Jump$tart Coalition for Personal Financial Literacy 2017)

Future Value

Years ago, there was such a thing as "penny candy," but does anything cost a penny anymore? This very question has led to a debate on whether the United States should even keep using this denomination of currency. Canada stopped producing pennies in 2012, so why does the U.S. keep minting a coin that costs more to make than it is worth?

According to a 2014 poll, more than two-thirds of Americans believe the penny still has value. Almost three-fourths of those surveyed believe that, by eliminating the penny, merchants will round up, thus causing Americans a greater expense (Americans for Common Cents n.d.). While they are still around, let us take advantage of the possibilities that pennies provide to students both as a foundation for learning financial literacy and as a means for practicing the basic mathematical skill of counting.

The following sections describe the mathematical and financial literacy concepts that this chapter will encompass, as well as the knowledge and experiences that elementary school students may bring to the lesson and take away on its completion. This chapter is not meant to be taught in one day, but rather over a series of days based on time availability.

Mathematics

This chapter focuses on many of the counting and cardinality (K.CC.A.3; K.CC.B.4) standards found in the kindergarten section of the Common Core State Standards for Mathematics. In part 1 of the lesson, students count a collection of pennies in varying configurations (K.CC.B.5). The teacher emphasizes conservation skills by maintaining the set of five pennies, yet changing the grouping arrangements. The students are then asked to write the numerals representing the count of pennies. They have the opportunity not only to correspond count to numbers but also to practice the correct symbols representing each numeral.

In part 2, students use their individual collection of pennies to compare value (K.CC.C.6) with other students. They will be able to compare the number of objects in a set, as well as the written numerals. Students will begin to see that there is order of magnitude to numbers and that they can be arranged in ascending or descending sequence (MP.2). They will also develop their mathematical vocabulary with words such as *greater than, less than,* and *equal to.*

Did you know?
It is thought that Benjamin Franklin suggested the design for the first one-cent coin (Americans for Common Cents n.d.). The penny was originally a lot heavier than it is today and made of pure copper, while it now consists of copper and zinc (United States Mint 2017). This metal material has helped make the penny into perhaps the most easily recognizable coins for children.

Part 3 of the lesson is an extension of part 2, in which students determine if they have enough money to "purchase" a good. Students trade money for items, which provides further practice with mathematical vocabulary, number order, and the concept of *equal.*

Financial Literacy

Students are fascinated by money, and it is never too early to start exposing them to the lifelong skills of responsibility and decision making based in financial literacy.

Across these activities, students use money. First, they must count out an appropriate payment (in pennies). They then make financial decisions by trading money for goods and make choices of which items can be purchased based on the amount of money in their pockets.

Student Knowledge and Experience

While kindergarten students might not have the cognitive readiness to understand the more abstract aspects of financial literacy, they probably have had previous exposure to coins, specifically the penny. Students probably also come to school with a sense of trading money for goods. This chapter will develop the skills of counting using pennies and spending money to make a purchase; teachers should also be sure to make connections to other relevant life experiences, such as buying lunch or food at the grocery store or in the cafeteria.

Lesson Plan

Learning Targets

- Count a collection of pennies.
- Compare sets of pennies.
- Make financial decisions on how to use pennies to purchase goods.

Resources and Tools

- **Resource 2.1**: Pocket Templates
- **Resource 2.2**: Toy Store
- Pennies (real or representations), one to ten per student
- *Pennies* by Mary Hill (2005) (or another book on pennies)
- *Alfie the Alligator* by Sandy Turley (2008) (or another book on comparisons)

Preparation Prior to Lesson

Part 1: Obtain the book *Pennies* by Mary Hill, or another book introducing pennies. Display five pennies on a projector. Prepare one copy of Resource 2.1, or just use a pocket on a piece of clothing.

Part 2: Obtain the book *Alfie the Alligator* by Sandy Turley, or another book on comparisons. Laminate enough copies of Resource 2.1 for each student. Place one to ten pennies in each pocket.

Part 3: Set up a classroom store with items priced between 1 and 10 cents, or display Resource 2.2 on the board so that all can see.

Children's Literature Option

For other resources besides the two books mentioned above, see the Literature for Financial Literacy list on p. 203.

Key Language

cent: One hundredth of a dollar in the U.S. money system. (It takes 100 cents to make one dollar.)

equal to: The first and second number are the same.

greater than: The first number is larger than the second number.

less than: The first number is smaller than the second number.

penny: A U.S. coin valued at one cent.

purchase: The act of trading money for goods or services.

value: How much something is worth in money.

Money Talks

Begin the lesson with a whole-class conversation about students' previous knowledge of money. Ask the following questions:

- What are some uses for money?
- What does it mean to "trade" money for things?
- Why might you want to spend money?
- Why might you want to save money?
- What things do you think about when deciding if you are going to buy something?

Sales Pitch

Say this popular poem to students, and then invite them to join you in saying it:

Penny, penny,
Easily spent.
Copper brown and worth one **cent.**

Ask students: "What does the poem tell us about pennies?"

Say the poem again, slowly. Ask students to try to think of one thing the poem tells us about pennies. Give students time to whisper their idea to a partner and then share out the ideas with the class. Next, with your students, read aloud the book *Pennies* by Mary Hill.

The Fine Print

Part 1: Counting

Tell students that when they have a small amount of money that they are responsible for keeping safe, they could place it in their pocket. Tell them that you found some money in your pocket this morning. Display five pennies with a fair amount of distance between each one. Point to each **penny** and ask for its **value**. Make sure students are aware that each penny is worth only one cent.

Next, push the coins closer together in a row. Have the students count the collection of pennies while you point to each one. To ensure one-to-one correspondence between the quantity of pennies and the number, you can draw an arrow down from each penny and write the numeral underneath (fig. 2.1).

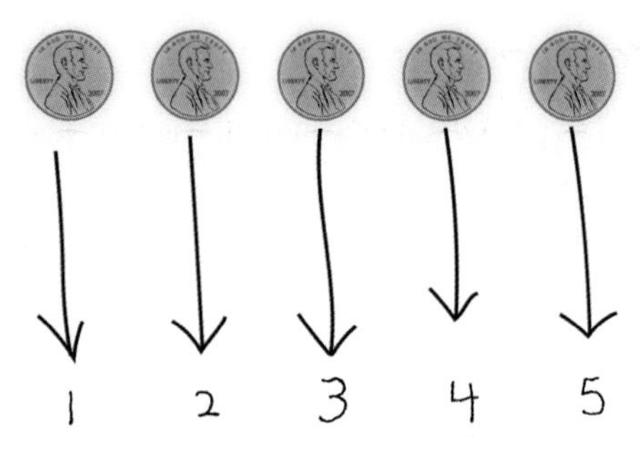

Fig. 2.1. Pennies and quantities

Now, push the pennies into a grouped set. Ask the students, "Do we still have the same amount of pennies?" Have the class count the group of pennies aloud. Repeat this step to ensure students recognize the conservation of the set and that no matter how they are configured, we still have the same number of pennies from our pockets (fig. 2.2).

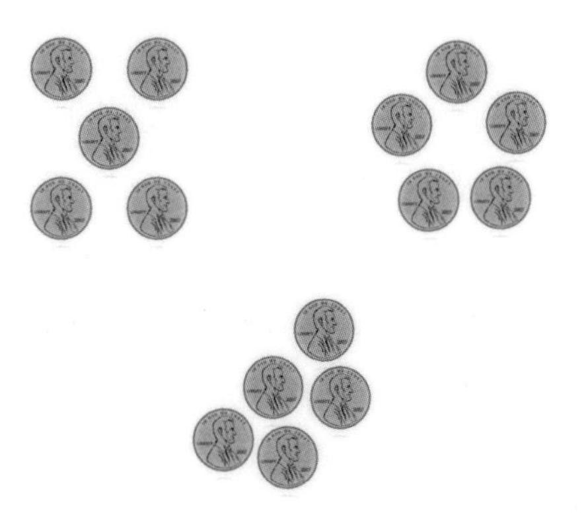

Fig. 2.2. Pennies in groups

As a class, determine the total amount of coins in the set, or cardinality of the set. Write the numeral and add the cents sign (fig. 2.3). Explain to the students that this is a symbol that we use to represent the word "cents" and that the amounts we will be counting today will all be written with this symbol.

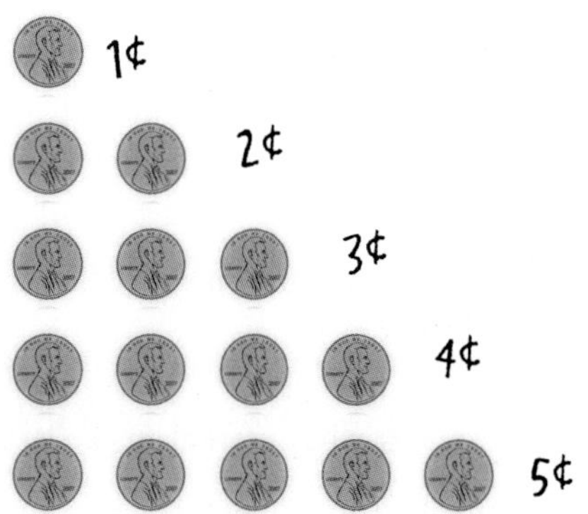

Fig. 2.3. Pennies counted in groups

Standards for Mathematical Practice: Students develop MP.2 when they use quantitative reasoning to identify the amount of money in the set (or pocket).

Use **Resource 2.1: Pocket Templates**, or your actual pocket, to ask students to guess how many pennies you have in your pocket. Talk about appropriate estimations. Then, turn your pocket inside out, or turn the paper pocket over, to show that you don't have any coins in your pocket. Ask the following questions:

- Now how many pennies do I have in my pocket? [zero]
- How much would this be in cents? [zero cents]

Write the numeral for zero, and explain to students that this is the number representing a count of no objects. Then write ¢ symbol representing zero cents.

Part 2: Comparing

Read *Alfie the Alligator* by Sandy Turley, or another picture book about comparing numbers, to the students. Explain to them that for this part of the activity, we are going to compare the amount of money we have in our pockets.

Standards for Mathematical Practice: Connecting numeral and quantity is a skill that needs to be repeated throughout the kindergarten mathematics curriculum and must be mastered before students are able to compare quantities and reason quantitatively (MP.2).

Give each student a laminated pocket that contains a collection of pennies (Resource 2.1). The pockets will include an amount between 1 cent and 10 cents. Have the students place their pennies on their desks and count them. Next, have the students use a dry-erase marker or a pencil to correctly record the numeral and symbol representing the value of the collection on a whiteboard or a piece of paper. Explain to the students that they are then to work with a partner and determine if the amount of money they have is **greater than, less than,** or **equal to** the amount of money their classmate does. Ask questions such as the following:

- How did you decide together who has more?
- What strategies did you use when deciding on this conclusion?

Repeat this activity several times so that students have ample practice with counting and comparing various quantities.

Part 3: Spending

Set up a classroom store with items priced between 1 and 10 cents, or display **Resource 2.2: Toy Store**. Give each student a laminated pocket that contains a collection of pennies (Resource 2.1). The pockets will include an amount between 1 cent and 10 cents. Have the students place their pennies on their desk and count them. Have them record the amount of money they have on a piece of paper. Give students the opportunity to shop in the toy store by looking at the items and prices only. Remind them that in order to **purchase** a good, they must have enough money. Refer back to part 2 and the discussion about having a greater, less, or equal amount of money when compared to their partner. This time the students are comparing their amount of money to the price of their item of choice.

Have students talk with others to discuss the possibilities. Listen for conversation about buying multiple items and about items that are more expensive or less expensive than their pocketful of pennies. Bring up interesting noticings and wonderings to the whole group. Ask students the following questions:

- What items can you buy with the amount of money you have?
- What items can't you buy, and why not?
- Could you buy more than one item?
- What happens if the item you buy costs less than the amount of money you have?

Closing the Deal

Ask the students this question: "If you combined your pennies with a partner, would you have more money, less money, or the same?" This serves as a precursor to addition. Have the students combine their pennies with their partners' pennies and count the amount again. Talk about their predictions of comparisons and whether they were correct. Now have the students "shop" again using their new amount. Make observations on whether the partners would go for one bigger item using the total amount of pennies, whether they would split the money evenly, or whether they would buy two objects or more. Discuss the strategies as a whole group, and listen to students' reasoning for the choices they made.

Getting Your Money's Worth

This activity can be modified for multiple grade levels by using various collections of coins and amounts. The level of difficulty can be adjusted based on the combination of coins and the value of the goods.

References and Resources

Americans for Common Cents (n.d.). "A Brief History of the U.S. Cent." http://www.pennies.org/index.php/penny-history.

Americans for Common Cents (n.d.). "More Than Two-Thirds of Americans Still Favor Keeping the Penny." http://www.pennies.org/index.php/penny-news/134-more-than-two-thirds-of-americans-still-favor-keeping-the-penny.

Hill, Mary. *Pennies.* New York: Rosen Book Works, Inc., 2005.

Jump$tart Coalition for Personal Financial Literacy. *National Standards in K–12 Personal Finance Education.* Fourth Edition. Second Printing. Washington, D.C.: Jump$tart Coalition for Personal Financial Literacy, 2017.

National Governors Association Center for Best Practices and Council of Chief State School Officers (NGA Center and CCSSO). *Common Core State Standards for Mathematics.* Washington, D.C.: NGA Center and CCSSO, 2010.

Turley, Sandy. *Alfie the Alligator: A Teaching Rhyme about Comparing Numbers.* Cucamonga, Calif.: TSA Worldwide Press, 2008.

United States Mint. "Circulating Coins" (2017). https://www.usmint.gov/learn/coin-and-medal-programs/circulating-coins?action=circPenny.

Resource 2.1

Pocket Templates

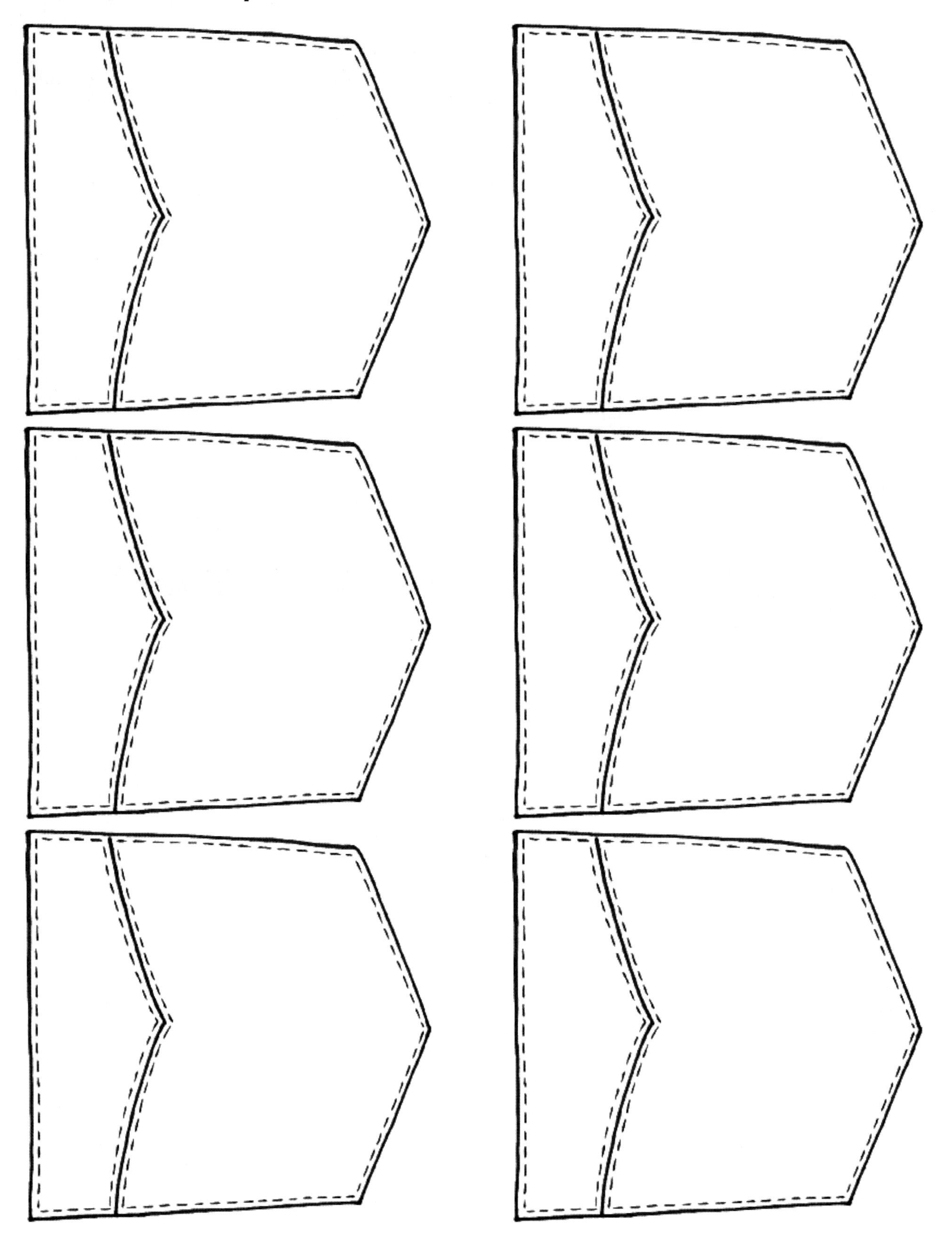

Resource 2.2

Toy Store

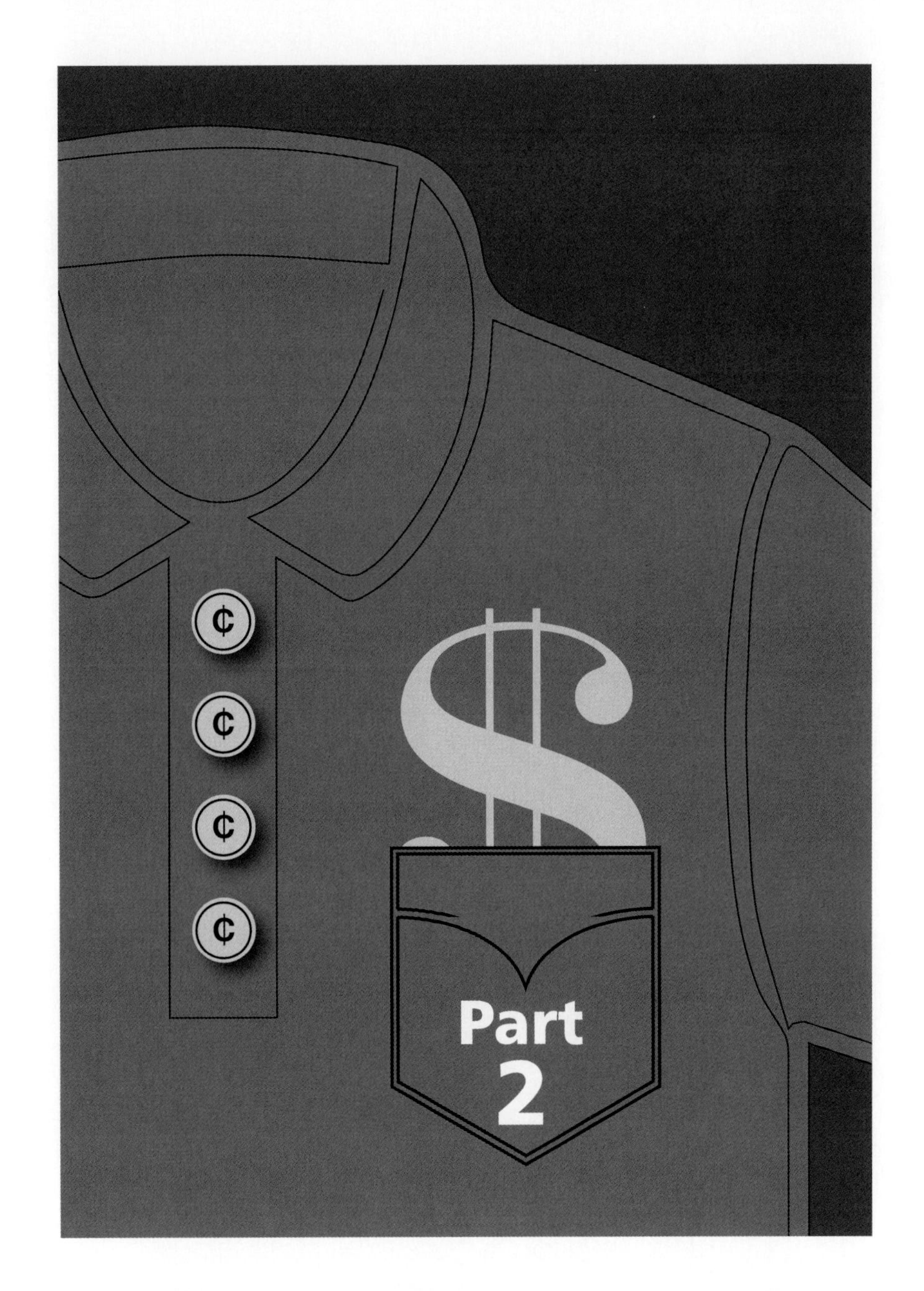

Operations and Algebraic Thinking

Chapter

Uncovering Hidden Funds

The Bottom Line

It is important to keep track of personal belongings and know where they are safely located.

Students explore ordinal numbers, counting by ones, skip-counting by fives and tens, and finding the missing value using pennies, nickels, and dimes.

Standards Met in This Chapter

CCSSM Standards and Practices

K.OA.A.1: Represent addition and subtraction with objects, fingers, mental images, drawings, sounds (e.g., claps), acting out situations, verbal explanations, expressions, or equations.

1.OA.B.4: Understand subtraction as an unknown-addend problem.

2.NBT.A.2: Count within 1000; skip-count by 5s, 10s, and 100s.

MP.3: Construct viable arguments and critique the reasoning of others.

MP.4: Model with mathematics.

(National Governors Association Center for Best Practices and Council of Chief State School Officers [NGA Center and CCSSO] 2010)

Jump$tart Standards

Risk Management and Insurance: Apply appropriate and cost-effective risk management strategies.

Knowledge Statements:

Overall: There are ways to keep possessions, including money, safely at home and other places.

(Jump$tart Coalition for Personal Financial Literacy 2017)

Future Value

It is important to know how much money you have and where you have placed it for safekeeping. Being responsible with your money and maintaining accurate records ensure that you do not have to search for hidden funds in the future!

The following sections describe the mathematical and financial literacy concepts that this lesson will encompass, as well as the knowledge and experiences that elementary students may bring to the lesson and take from it on its completion.

Mathematics

This chapter spans a band of grade levels (from kindergarten to grade 2), providing a tiered approach to skip-counting (2.NBT.A.2), addition and subtraction (K.OA.A.1), or finding a missing addend (1.OA.B.4). Students will approach the problems differently, and all logical pathways should be acknowledged and shared (MP.3). Give students ample time for discussion and for comparing the different ways that students represent the situations symbolically and connect the situations to the symbols (MP.4).

Financial Literacy

If someone owes you money, and you do not collect it, that is considered unclaimed funds. These funds can be a result of an overpayment to a credit card, a tax refund, an insurance claim, or unpaid wages. "Hidden" money can also be money you have but you do not see. For instance, we keep money in financial institutions for safety. We know that it is there and protected, even when we cannot see it.

As part of this chapter, it is important to address with students the financial literacy skills of coin identification and value, which leads into a discussion regarding responsibility and keeping track of one's money. Students will be exposed to instances when they might have to count coins or calculate a missing addend. Remind students that it is important for us to know how much money we have and where it is located so that funds are not temporarily or permanently lost.

Student Knowledge and Experience

Students may have limited experience with depositing money into banks or other financial institutions. Because of this, the thought of having money that you cannot see might be unfamiliar. Start with the notion of "hidden money" being located in a piggy bank or other safe space, where it might not be seen but is still available.

Students who have not had previous experience with formal algebraic methods will create their own strategies for finding out how much money is hidden, as well as pick up new strategies from classroom discussion. Encourage students to verbalize, draw a picture, or write a number sentence to illustrate their mathematical model for finding the hidden part. As students become more familiar with the skill, you could manipulate the money strips to include a variety of coins, patterns, and amounts that can lead to the discussion of multiple variables or how one variable does not always mean one solution.

Did you know?
As of 2017, there is not one centralized location for finding unclaimed funds nationally. To learn more about unclaimed funds, begin with your state, as well as other sources found at https://www.usa.gov/unclaimed-money (USA.gov 2017).

Lesson Plan

Learning Targets

- Use mathematical reasoning strategies to find the amount of hidden money.
- Tell why we might need to calculate the missing amount.
- Describe ways to keep money safe.

Resources and Tools

- **Activity Sheet 3.1a:** Modeling with Money—Pennies
- **Activity Sheet 3.1b:** Modeling with Money—Nickels
- **Activity Sheet 3.1c:** Modeling with Money—Dimes
- **Resource 3.1:** Coin Strips
- **Resource 3.2:** Cover Strips
- **Resource 3.3:** Clue Templates
- Coin wrappers
- Coins (real or not), ten or fewer per student
- *Counting with Common Cents: Penny's Journey* by Deirdre McCarthy (2014) or *The Coin Counting Book* by Rozanne Lanczak Williams (2001) (or another book about counting a collection of coins)

Preparation Prior to Lesson

Part 1: Obtain *Counting with Common Cents: Penny's Journey* by Deirdre McCarthy, *The Coin Counting Book* by Rozanne Lanczak Williams, or another book to review counting coins. Obtain coin wrappers to show students. Prepare strips of coins (Resource 3.1) or actual coins laid out for a projector along with the cover strips in Resource 3.2.

Part 2: Prepare ten or fewer coins for each student. Provide each student with one or more clue templates from Resource 3.3. Activity Sheet 3.1a, b, or c can be used as an assessment. Make one copy for each student.

Children's Literature Option

See the Literature for Financial Literacy list on p. 203 for additional books to consider incorporating into the lesson.

Key Language

addend: A number that is added to another number.

funds: A sum of money made available or saved for a particular purpose.

model: Using pictures, words, or symbols to represent a mathematical situation.

total: The whole amount.

Money Talks

Discuss with students whether they have ever found money and where it was located. (For example, they might have found a coin on the ground or in a pocket of a coat.) Talk about these "unclaimed **funds**" and the steps that should be taken, if necessary, to find the rightful owner. Explain that the "finders keepers" rule does not always apply and when and how we should return lost money.

Explain to students that sometimes we have money that we cannot see. For example, if we deposit money into the bank, we don't see the actual bills and coins, but we know that our money is there and is safe. One way we deposit large amount of coins is by using paper coin wrappers. Show students the various coin wrappers and ask the following questions:

- Do you know how many pennies, nickels, dimes, or quarters are in a coin roll? (Make sure to bring in actual coin wrappers to show the students. They may notice the total printed on the outside.)
- We know the total printed on the outside, but how many coins should we include inside the wrapping?

Discuss with students how banks can verify that the roll contains the correct amount. Ask:

- If I know that 50 dimes go in the wrapper, but I only have 40 dimes, how many more dimes will I need to complete the roll?

Being able to reason mathematically to find missing parts or totals can help students solve real-world application problems related to money and strengthen their skills in financial literacy.

Sales Pitch

Standards for Mathematical Practice: Students develop MP.3 by making conjectures regarding the missing money and communicate their reasoning to others.

Begin the lesson with a literature connection using *Counting with Common Cents: Penny's Journey* by Deirdre McCarthy or *The Coin Counting Book* by Rozanne Lanczak Williams to activate students' prior knowledge about counting money.

The Fine Print

Part 1: Missing Parts

Show the students a strip of coins without giving them enough time to count them (**Resource 3.1: Coin Strips**). Explain to the students that you will be partially covering the strip of coins and that they will only be able to see a part of the total money (Resource 3.1). It will be the students' job to figure out how many coins are on the whole strip. (This activity can also be done using an interactive board or PowerPoint technology.) Clarify for the students that they will accomplish this by listening to clues and using their mathematical thinking to find the missing part.

Here we share three examples (related to figs. 3.1, 3.2, and 3.3) with a series of questions scaffolded to reason about the missing number of coins and reason about the missing value of coins. Depending on the experiences of the students, you may choose to omit some of the questions or keep and add more structure or visuals to the lesson. After a clue is given, you can ask the students to explain their reasoning and justify their answer.

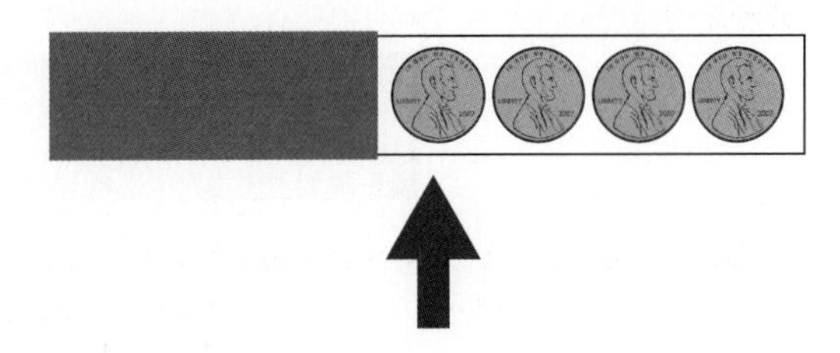

Fig. 3.1. First hidden coins example

Clue: The arrow is pointing at the fourth penny (fig. 3.1).

Possible questions for the students include—

- How many pennies are hidden?
- How many total pennies are there?
- What equation can we write to describe the visible coins, hidden coins, and total coins? Repeat the questions for the value of the coins.
- Finally, what are some equations that describe the situation? [Generate a lot of ideas.]

Standards for Mathematical Practice: Students develop MP.4 as they consider different ways to write equations to describe the quantity of seen and hidden coins.

Fig. 3.2. Second hidden coins example

Clue: The arrow is pointing at the second nickel from the left, or sixth nickel from the right (fig. 3.2).

Possible questions for the students include—

- How many nickels are hidden?
- What number will we have to skip-count by to find the total value of the coins?
- What equation can we write to describe the visible coins, hidden coins, and total coins? Repeat the questions for the value of the coins.
- Finally, what are some equations that describe the situation? [Generate a lot of ideas.]

Fig. 3.3. Third hidden coins example

Clue: The arrow is pointing at the sixth dime (fig. 3.3).

Possible questions for the students include—

- How many dimes are hidden?
- What number will we have to skip-count by to find the total value of the coins?
- What equation can we write to describe the visible coins, hidden coins, and total coins? Repeat the questions for the value of the coins.
- Finally, what are some equations to describe the situation? [Generate a lot of ideas.]

Bring up interesting noticings and wonderings to the whole group.

Part 2: Student-Generated Scenarios

Next, give the students the opportunity to make up their own hidden money task for a partner. Pass out ten or fewer coins of whatever denomination you choose. Ask the students to place the coins in a row and cover part of them with a piece of paper or **Resource 3.2: Cover Strips.** Have them verbalize or write down one or more clues that they might give their partner on how many coins are hidden. This can be done on a piece of paper or on a section handed out from **Resource 3.3: Clue Templates**. Give students the freedom to explore this problem in various ways, such as using ordinal words, skip-counting, finding a missing **addend**, or calculating a **total**. Have the students try their clues on a partner. Discuss the strategies students used while performing the task as a whole group. Repeat the activity several times to give students the opportunity to work with different amounts of money. Observe how students come up with their clues and the strategies they use to solve their partner's task. Ask students to explain their thoughts and justify their reasoning.

Closing the Deal

End the lesson with a discussion about the importance of keeping money safe. Ask the following questions:

- How might you keep track of your money?
- How can you keep your money safe?

Whether they place it in a piggy bank, an institutional bank, a credit union, or just in their pocket, they need to be aware of the money they have and where it is located. Students should also be cognizant of missing addends when they are working toward a goal or total. Using the strategies found in this chapter, students will be able to reason and calculate the "hidden" amount. Being responsible for one's money is essential so that it does not become lost or unclaimed funds.

Finally, give students the opportunity for independent practice and to **model** the mathematics using **Activity Sheet 3.1a, b, or c.** These sheets can be used as a formal assessment of the chapter.

Getting Your Money's Worth

This activity can be modified for multiple grade levels by using various dominations of coins and amounts. The level of difficulty can be adjusted and the emphasis on counting or algebra can be prioritized depending on student ability.

References and Resources

Jump$tart Coalition for Personal Financial Literacy. *National Standards in K–12 Personal Finance Education*. Fourth Edition. Second Printing. Washington, D.C.: Jump$tart Coalition for Personal Financial Literacy, 2017.

McCarthy, Deirdre. *Counting with Common Cents: Penny's Journey*. North Charleston, S.C.: CreateSpace Independent Publishing Platform, 2014.

National Governors Association Center for Best Practices and Council of Chief State School Officers (NGA Center and CCSSO). *Common Core State Standards for Mathematics*. Washington, D.C.: NGA Center and CCSSO, 2010.

USA.gov. "Unclaimed Money from the Government" (2017). https://www.usa.gov/unclaimed-money.

Williams, Rozanne Lanczak. *The Coin Counting Book*. Watertown, Mass.: Charlesbridge, 2001.

Activity Sheet 3.1a

Modeling with Money—Pennies

Create the following models using drawings, stickers, or stamps to represent the coins.

1. This strip of coins is equal to 7¢. How many pennies are hidden?

Draw your model in the box.

2. This strip of coins is equal to 10¢. How many pennies are hidden?

Draw your model in the box.

Activity Sheet 3.1a—*page 2*

Modeling with Money—Pennies

3. This strip of coins is equal to 6¢. How many pennies are hidden?

Draw your model in the box.

4. This strip of coins is missing 3¢. How many pennies are there in total?

Draw your model in the box.

Activity Sheet 3.1b

Modeling with Money—Nickels

Create the following models using drawings, stickers, or stamps to represent the coins.

1. This strip of coins is equal to 30¢. How many nickels are hidden?

Draw your model in the box.

2. This strip of coins is equal to 50¢. How many nickels are hidden?

Draw your model in the box.

Activity Sheet 3.1b—*page 2*

Modeling with Money—Nickels

3. This strip of coins is equal to 35¢. How many nickels are hidden?

Draw your model in the box.

4. This strip of coins is missing 15¢. How many nickels are there in total?

Draw your model in the box.

Activity Sheet 3.1c

Modeling with Money—Dimes

Create the following model using drawings, stickers, or stamps to represent the coins.

1. This strip of coins is equal to 80¢. How many dimes are hidden?

Draw your model in the box.

2. This strip of coins is equal to $1.00. How many dimes are hidden?

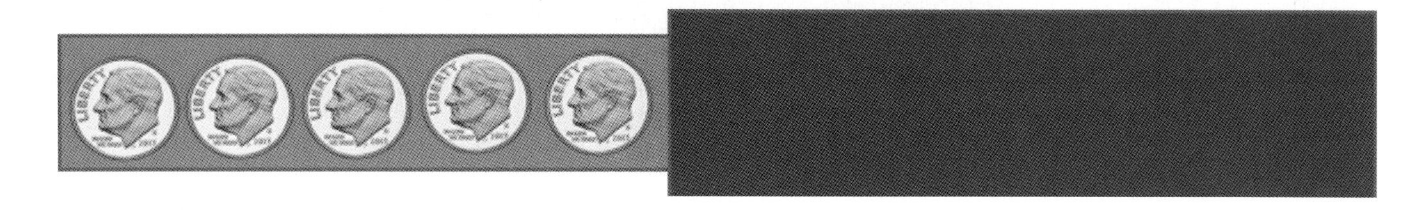

Draw your model in the box.

Activity Sheet 3.1c—*page 2*

Modeling with Money—Dimes

3. This strip of coins is equal to 90¢. How many dimes are hidden?

Draw your model in the box.

4. This strip of coins is missing 40¢. How many dimes are there in total?

Draw your model in the box.

Resource 3.1

Coin Strips

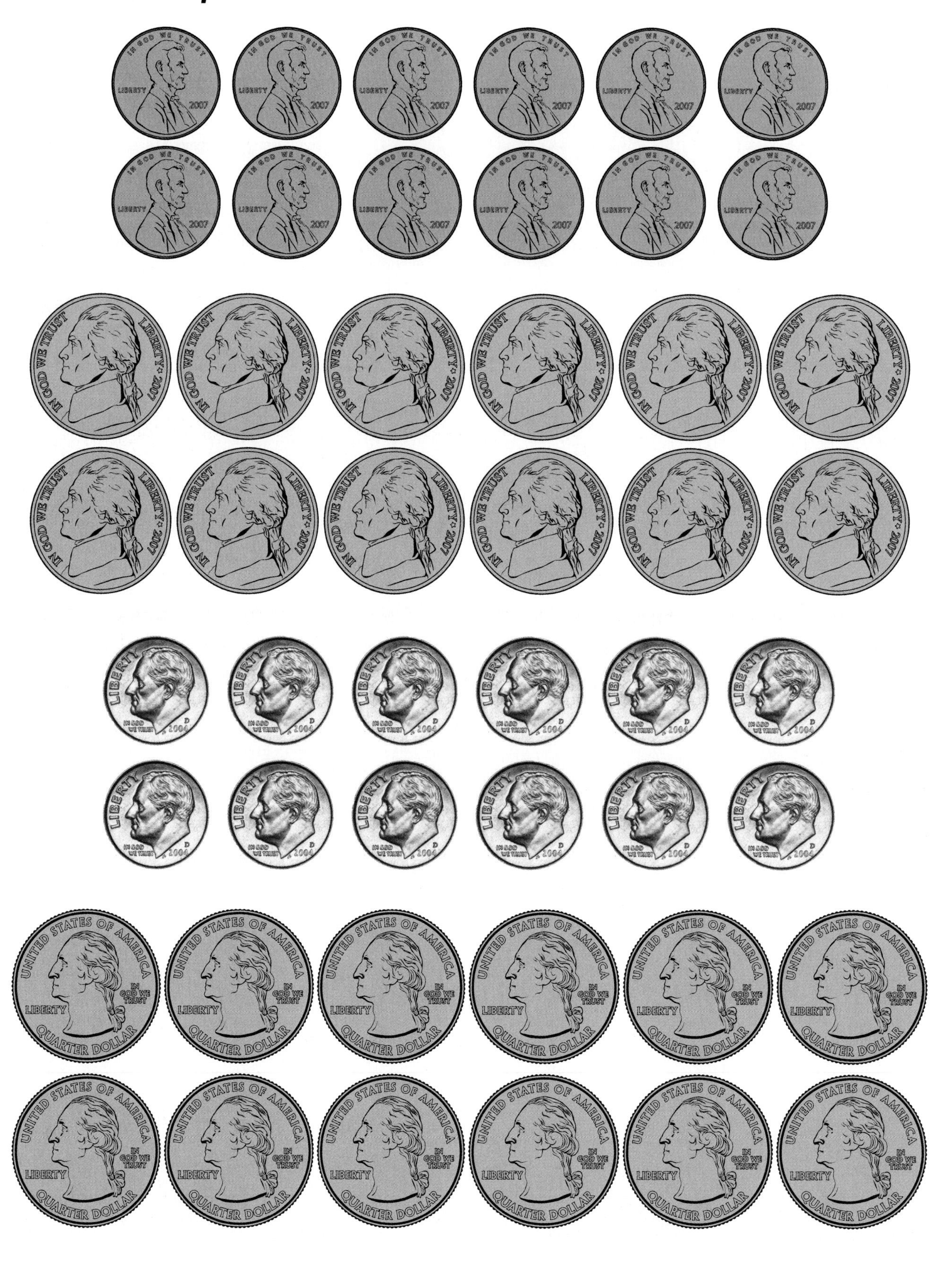

Resource 3.2

Cover Strips

HIDDEN FUNDS

HIDDEN FUNDS

HIDDEN FUNDS

HIDDEN FUNDS

HIDDEN FUNDS

HIDDEN FUNDS

HIDDEN FUNDS

Resource 3.3

Clue Templates

CLUE

CLUE

CLUE

CLUE

CLUE

CLUE

Chapter

I Got a Dollar!

The Bottom Line

There are numerous ways to make a dollar using pennies, nickels, dimes, and quarters.

Students use algebraic reasoning and properties of addition to explore and verify the different combinations of candy that total a dollar.

Standards Met in This Chapter

CCSSM Standards and Practices

K.CC.A.1: Count to 100 by ones and by tens.

K.OA.A.2: Solve addition and subtraction word problems, and add and subtract within 10, e.g., by using objects or drawings to represent the problem.

1.OA.A.1: Use addition and subtraction within 20 to solve word problems involving situations of adding to, taking from, putting together, taking apart, and comparing, with unknowns in all positions, e.g., by using objects, drawings, and equations with a symbol for the unknown number to represent the problem.

1.OA.B.3: Apply properties of operations as strategies to add and subtract. *Examples: If $8 + 3 = 11$ is known, then $3 + 8 = 11$ is also known. (Commutative property of addition.) To add $2 + 6 + 4$, the second two numbers can be added to make a ten, so $2 + 6 + 4 = 2 + 10 = 12$. (Associative property of addition.)*

1.NBT.C.4: Add within 100, including adding a two-digit number and a one-digit number, and adding a two-digit number and a multiple of 10, using concrete models or drawings and strategies based on place value, properties of operations, and/or the relationship between addition and subtraction; relate the strategy to a written method and explain the reasoning used. Understand that in adding two-digit numbers, one adds tens and tens, ones and ones; and sometimes it is necessary to compose a ten.

2.OA.A.1: Use addition and subtraction within 100 to solve one- and two-step word problems involving situations of adding to, taking from, putting together, taking apart, and comparing, with unknowns in all positions, e.g., by using drawings and equations with a symbol for the unknown number to represent the problem.

2.NBT.A.2: Count within 1000; skip-count by 5s, 10s, and 100s.

2.NBT.B.7: Add and subtract within 1000, using concrete models or drawings and strategies based on place value, properties of operations, and/or the relationship between addition and subtraction; relate the strategy to a written method. Understand that in adding or subtracting three-digit numbers, one adds or subtracts hundreds and hundreds, tens and tens, ones and ones; and sometimes it is necessary to compose or decompose tens or hundreds.

2.MD.C.8: Solve word problems involving dollar bills, quarters, dimes, nickels, and pennies, using $ and ¢ symbols appropriately. Example: If you have 2 dimes and 3 pennies, how many cents do you have?

3.NBT.A.2: Fluently add and subtract within 1000 using strategies and algorithms based on place value, properties of operations, and/or the relationship between addition and subtraction.

MP.4: Model with mathematics.

MP.7: Look for and make use of structure.

(National Governors Association Center for Best Practices and Council of Chief State School Officers [NGA Center and CCSSO] 2010)

Jump$tart Standards

Spending and Saving: Apply strategies to monitor income and expenses, plan for spending and save for future goals.

Standard 3. Describe how to use different payment methods.

Knowledge Statements:

Overall:

- People trade money to buy goods or services (tasks performed by others).
- Different goods or services have different prices.

Standard 3:

- Verify the total cost of a purchase that includes multiple items.

Financial Decision Making: Apply reliable information and systematic decision making to personal financial decisions.

Standard 4. Make criterion-based financial decisions by systematically considering alternatives and consequences.

Knowledge Statements:

Overall:

Money can be spent only once—after buying something, a person needs more money to buy something else.

Explain how limited personal financial resources affect the choices people make.

(Jump$tart Coalition for Personal Financial Literacy 2017)

Future Value

There are many ways to use coins to total a dollar. Students will need to strengthen these skills as they progress to larger total amounts when making a purchase, which will also be reinforced through practice with mathematical operations. By understanding that various combinations can create a dollar, students are practicing strategies that will later aid in the process of making change.

The following sections describe the mathematical and financial literacy concepts that this lesson will encompass, as well as the knowledge and experiences that elementary students may bring to the lesson and take from it on its completion.

Did you know?
We think of one dollar as being valued the same now as in the past, but inflation has played a role in determining its worth. According to the Bureau of Labor Statistics, a dollar in 1916 has the same buying power as $22.18 in 2016! (Bureau of Labor Statistics 2016).

Mathematics

The lesson in this chapter has the potential to cover a variety of mathematics topics, depending on the grade level. The chapter's overarching themes are the properties of addition, promoting algebraic reasoning (MP.4), and problem solving (K.CC.A.1; K.OA.A.2; 1.OA.A.1; 1.OA.B.3; 1.NBT.C.4; 2.OA.A.1; 2.NBT.A.2; 2.NBT.B.7; 3.NBT.A.2). Students will need to use quantitative reasoning skills to create an organized list, which will ensure they have found all combinations (MP.7) of one dollar that are possible (2.MD.C.8). This chapter could be extended, however, to include multiplication, percentages, decimals, and rates, if desired.

Financial Literacy

In this chapter, students will continue to review coin identification and value, as well as counting a small collection of coins. They will also explore how money is not limitless and that people trade money for goods or services. Students need to be aware that different goods can have different prices and that sometimes we must make a choice on how we want to spend our money. These choices can be based on personal values as well as on economic reasoning. When making a purchase of multiple items, students must be able to verify the total cost. This is a life skill that they will use when grocery shopping or in other purchasing situations.

Student Knowledge and Experience

Some students may have previous experience with spending money or purchasing candy. This chapter will build on these experiences during discussions where students have had the opportunity to purchase something (e.g., bought a gift for a friend or family member, purchased food at a concession stand, or put a quarter into a gumball machine). An important concept in this chapter is that money is not limitless—when it is gone, it is gone. Because it is not limitless, we sometimes have to make difficult decisions on how to spend our money. This can also lead to further discussion on how we don't have to spend money immediately when we receive it, but rather we can save for a future purchase or something that might be more expensive.

Lesson Plan

Learning Targets

- Use coins to create a variety of ways to total one dollar.
- Purchase and verify multiple items totaling one dollar.
- Encourage the use of the properties of addition to explore combinations.
- Promote algebraic reasoning when creating possible combinations that total a dollar.

Resources and Tools

- **Activity Sheet 4.1a:** I Got a Dollar! (without coin amounts), one per student
- **Activity Sheet 4.1b:** I Got a Dollar! (with coin amounts), one per student
- **Activity Sheet 4.2:** I Got a Dollar!–Answer Key
- **Resource 4.1a:** I Got a Dollar! word problem (without coin amounts)
- **Resource 4.1b:** I Got a Dollar! word problem (with coin amounts)

- Candy (more than needed to make a dollar—e.g., 6 lollipops, 12 pieces of bubble gum, and 22 chocolate pieces, one bag per student or pair of students)
- *A Dollar, a Penny, How Much and How Many?* by Brian P. Cleary (2014) (or another book on various ways to create a dollar)
- *Daphney Dollar's First Trip to the Store* by Sharon M. Lewis (2016) (or another book relating to allowance and shopping)

Preparation Prior to Lesson

Part 1: Obtain a copy of *A Dollar, a Penny, How Much and How Many?* by Brian P. Cleary, obtain a copy of *Daphney Dollar's First Trip to the Store* by Sharon M. Lewis

Part 2: Project Resource 4.1a or 4.1b for all students to see, prepare enough copies of Activity Sheet 4.1a or 4.1b as needed for each student, prepare bags of candy (more than needed to make a dollar, such as 6 lollipops, 12 pieces of bubble gum, and 22 chocolate pieces, with one bag per student or pair of students)

Children's Literature Option

See the Literature for Financial Literacy list on p. 203 for additional related children's books.

Key Language

additive identity property: The sum of any real number and zero is the original number.

allowance: An amount of money regularly given to a person for a purpose.

associative property of addition: When adding three or more addends, the sum is always the same, regardless of how they are grouped.

commutative property of addition: Addends can be added in any order and still result in the same sum.

overspend: Spending more than was planned.

Money Talks

Ask children if they have ever received money as a gift from someone. Discuss what they did with that money (e.g., saved it, spent it, lost it). Talk with them about their decision-making process when making a purchase. Ask the following questions:

- Do you go into the store with a set amount of money and plan to spend that whole amount?
- Have you ever used your money to buy something and then wished you still had it because you found something else you wanted to buy?
- What type of planning do you do before making a purchase?
- What are some reasoning skills you use before buying? (e.g., Is it better to get more, as in number of items, for your money, or more, as in value?)

Sales Pitch

Begin the lesson with a literature connection using *A Dollar, a Penny, How Much and How Many?* by Brian P. Cleary to get students thinking about the many ways you can make a dollar with coins or a bill. Review the various coins and their values, and give some examples of how to use these coins to total one dollar.

The Fine Print

Part 1: How Do We Get Money?

Did you know? According to a Gallup poll in 2013, only about one-third of Americans (32 percent) prepare and keep a household budget (Jacobe 2013).

Read *Daphney Dollar's First Trip to the Store by* Sharon M. Lewis. In this book, Daphney earns her allowance and takes her first trip to the store. Discuss with the students what the term **allowance** means. Ask them to share experiences with allowances if they have any. Ask students what are some other ways we get money (e.g., working, gifts, winnings). Talk with them about how money is not limitless and that when we go to the store, we usually need to have a plan of what we are going to buy so that we do not **overspend**. We also cannot spend money more than once. Once the money is spent, we will have to save more money to make another purchase.

Part 2: I Got a Dollar!

Explain to the students that they just received a dollar from their grandma for their birthday and they are going to shop at the candy store. Hand each pair or group of students a bag of candy that contains 6 lollipops, 12 pieces of bubble gum, and 22 chocolate pieces. (Feel free to substitute any candy and make sure you are aware of any student food allergies.) Display on the board (or supply students with **Resource 4.1a or 4.1b: I Got a Dollar!**) the following word problem:

> My grandma gave me a dollar for my birthday. I went to the candy store to buy myself some treats. Lollipops were a quarter each (25¢), bubble gum was a dime (10¢), and chocolate pieces were a nickel each (5¢). How many lollipops, bubble gum, and chocolate pieces could I buy?

Standards for Mathematical Practice: As students manipulate the candy and illustrate their number thinking, they are modeling with mathematics (MP.4).

Let students explore different methods for deciding what candy to purchase by encouraging them to model the mathematics by using the candy or by drawing coins. Have students illustrate their mathematical model using pictures, words, or symbols. After five to ten minutes, ask students to share what they decided to buy and how they made that decision. Discuss how different students will have different answers. If some did not choose to spend all their money, talk about why and what they will do with the change.

Part 3: Making an Organized List

Next, ask students how we could go about making a complete list of the different ways to spend the dollar. Pass out **Activity Sheet 4.1a or 4.1b: I Got a Dollar!** as appropriate. Provide students with the parameters that they must spend the whole dollar and that they cannot overspend. As the groups of students are working, individually ask them if they have discovered patterns or surprises. Talk with them about ways that they can organize their thinking into a list and how these patterns can help them find all possible combinations of treats. **Activity Sheet 4.2: I Got a Dollar!–Answer Key** demonstrates one way to make an organized list of answers, by working through combinations that have the maximum possible number of lollipops, and then the maximum number of bubble gums.

Standards for Mathematical Practice: Students are looking for and making use of structure by discovering patterns, such as two bubble gum pieces equals one chocolate piece, or two nickels equals one dime (MP.7).

Closing the Deal

Give the students ample time to compile a comprehensive list (Activity Sheet 4.1). Guide them toward an organized list to make sure they attempt every combination. Talk with the students about reversing the order of candy purchased—for example, buying the chocolate pieces before the lollipops—and discuss how the **commutative property of addition** tells us that it does not matter what order we add, it will result in the same sum.

Discuss the **associative property of addition** by grouping the nickels with the quarters and then adding on the dimes, or grouping the quarters and dimes and adding on the nickels. Give students the opportunity to discover that not only does order not matter, but grouping will not affect the sum either.

Finally, ask students if any of their candy columns had zero pieces of candy. Talk about what it means to add zero to another number and that the **additive identity property** of addition tells us that the sum of any number and zero is the original number.

Wrap up the lesson by asking the students the following questions:

- How did you decide what candy to purchase in the first part of the activity?
- How did you keep track of your purchases?
- What reasoning skills did you use when making your organized list?
- Did you notice any patterns?
- Would the price of the candy affect which pieces you would purchase?
- If you didn't have to spend the whole dollar, would you?
- What would you choose to do with the change?
- For older students, you could ask: What is a way that you could use multiplication to help you calculate the total cost of the candy?

Analyze how your students are starting to use algebraic reasoning to solve the word problem. Extend the activity for students who are ready to use multiplication or the distributive property as their strategies.

Getting Your Money's Worth

The level of difficulty for this task can be from kindergarten (using all pennies) to fifth grade (incorporating multiplication and higher denominations of money). Students who are ready for the more challenging mathematical skills can begin to look at the problem algebraically through variables. For example, $25l + 10b + 5c = 100$ can be used where l = lollipops, b = bubble gum, and c = chocolate pieces. Some students will be able to figure out the numerical coefficient (the set price of the candy), and encourage them to use a variable to represent the number of pieces of each type of candy.

References and Resources

Bureau of Labor Statistics (U.S. Department of Labor). "CPI Inflation Calculator" (2016). http://www.bls.gov/data/inflation_calculator.htm

Cleary, Brian P. *A Dollar, a Penny, How Much and How Many?* Minneapolis, Minn.: Lerner, 2014.

Jacobe, Dennis. "One in Three Americans Prepare a Detailed Household Budget" (2013). http://news.gallup.com/poll/162872/one-three-americans-prepare-detailed-household-budget.aspx.

Jump$tart Coalition for Personal Financial Literacy. *National Standards in K–12 Personal Finance Education*. Fourth Edition. Second Printing. Washington, D.C.: Jump$tart Coalition for Personal Financial Literacy, 2017.

Lewis, Sharon M. *Daphney Dollar's First Trip to the Store*. Bowie, Md.: Fiscal Pink, 2016.

National Governors Association Center for Best Practices and Council of Chief State School Officers (NGA Center and CCSSO). *Common Core State Standards for Mathematics*. Washington, D.C.: NGA Center and CCSSO, 2010.

Activity Sheet 4.1a

I Got a Dollar!

My grandma gave me a dollar for my birthday. I went to the candy store to buy myself some treats. Lollipops were a quarter, bubble gums a dime, and chocolate pieces a nickel. How many lollipops, bubble gums, and chocolate pieces could I buy?

lollipops	bubble gums	chocolate pieces

Activity Sheet 4.1b

I Got a Dollar!

My grandma gave me a dollar for my birthday. I went to the candy store to buy myself some treats. Lollipops were a quarter (25¢), bubble gums a dime (10¢), and chocolate pieces a nickel (5¢). How many lollipops, bubble gums, and chocolate pieces could I buy?

lollipops	bubble gums	chocolate pieces

Activity Sheet 4.2

I Got a Dollar!–Answer Key

My grandma gave me a dollar for my birthday. I went to the candy store to buy myself some treats. Lollipops were a quarter, bubble gums a dime, and chocolate pieces a nickel. How many lollipops, bubble gums, and chocolate pieces could I buy?

Lollipops	Bubble Gums	Chocolate Pieces
4	0	0
3	2	1
3	1	3
3	0	5
2	5	0
2	4	2
2	3	4
2	2	6
2	1	8
2	0	10
1	7	1
1	6	3
1	5	5
1	4	7
1	3	9
1	2	11
1	1	13
1	0	15
0	10	0
0	9	2
0	8	4
0	7	6
0	6	8
0	5	10
0	4	12
0	3	14
0	2	16
0	1	18
0	0	20

I Got a Dollar!

My grandma gave me a dollar for my birthday. I went to the candy store to buy myself some treats. Lollipops were a quarter, bubble gums a dime, and chocolate pieces a nickel. How many lollipops, bubble gums, and chocolate pieces could I buy?

Resource 4.1b

I Got a Dollar!

My grandma gave me a dollar for my birthday. I went to the candy store to buy myself some treats. Lollipops were a quarter **(25¢)**, bubble gums a dime **(10¢)**, and chocolate pieces a nickel **(5¢)**. How many lollipops, bubble gums, and chocolate pieces could I buy?

Chapter

Garage Sale!

The Bottom Line

Material objects have value, and students can buy and sell objects to meet their wants and needs.

In typical everyday transactions, you might use any combination of the four basic operations in mathematics: addition, subtraction, multiplication, and division. In this activity, students will explore using the four functions to purchase items at a garage sale. They will then decide what to do with the money they have left over.

Standards Met in This Chapter

CCSSM Standards and Practices

3.OA.C.7: Fluently multiply and divide within 100, using strategies such as the relationship between multiplication and division (e.g., knowing that $8 \times 5 = 40$, one knows $40 \div 5 = 8$) or properties of operations. By the end of Grade 3, know from memory all products of two one-digit numbers.

3.OA.D.8: Solve two-step word problems using the four operations. Represent these problems using equations with a letter standing for the unknown quantity. Assess the reasonableness of answers using mental computation and estimation strategies including rounding.

4.OA.A.3: Solve multistep word problems posed with whole numbers and having whole-number answers using the four operations, including problems in which remainders must be interpreted. Represent these problems using equations with a letter standing for the unknown quantity. Assess the reasonableness of answers using mental computation and estimation strategies including rounding.

MP.1: Make sense of problems and persevere in solving them.

MP.5: Use appropriate tools strategically.

MP.6: Attend to precision.

(National Governors Association Center for Best Practices and Council of Chief State School Officers [NGA Center and CCSSO] 2010)

Jump$tart Standards

Spending and Saving: Apply strategies to monitor income and expenses, plan for spending and save for future goals.

Standard 3. Describe how to use different payment methods.

Knowledge Statements:

Overall: Spending plan (budget) is a guide to help people balance money coming in (income) and money going out (expenses).

Standard 3: Verify the total cost of a purchase that includes multiple items.

Employment and Income: Use a career plan to develop personal income potential.

Standard 2: Compare sources of personal income and compensation.

Knowledge Statements:

Standard 2:

- Identify jobs children can do to earn money.
- Give examples of sources of income other than a wage or salary..

(Jump$tart Coalition for Personal Financial Literacy 2017)

Future Value

When shopping, for groceries or other items, students should be able to perform mental computations. Being able to mentally calculate and estimate using the four basic mathematical operations is an important life skill for a financially literate adult, especially when making purchases within a budget. Mental mathematics is a life skill that will come in handy when required to make quick financial decisions without the use of a calculator.

Additionally, students think in this lesson about items having value. In the future, students might want to have a garage sale of their own to make some money from items that they no longer want or need. As the saying goes: "One man's trash is another man's treasure." Holding a garage sale is one way to make income without a wage or salary, and it is also an excellent opportunity for students to apply learned mathematical skills such as using the four operations.

The following sections describe the mathematical and financial literacy concepts that this lesson will encompass, as well as the knowledge and experiences that elementary students may bring to the lesson and take from it on its completion.

Did you know? The second Saturday in August is known as National Garage Sale Day. Saturdays and summertime are a perfect combination for garage sales (nationaldaycalendar.com n.d.).

Mathematics

Throughout these activities, students will be performing mental calculations using the four basic mathematical operations (MP.1; MP.6). Frequent practice of these skills leads to students' fluency in mathematical operations (3.OA.C.7) and the ability to better make decisions based on comparisons (MP.5; 4.OA.A.3). Students who are fluent in mental mathematics (3.OA.D.8) can extend their thinking through budgeting and calculations to keep from overspending. Depending on the grade level, or for differentiating purposes, these activities can also be modified to include place value, rounding, estimation, and decimal practice.

Standards for Mathematical Practice: When performing their mental calculations, students will need to attend to precision (MP.6).

Financial Literacy

This chapter gives students the chance to experience real-life situations they might encounter in which they will need to use some basic math skills to help them make decisions. Before one makes a purchase, it is always a good idea to have a budget or plan. Students can take the skills learned in

Did you know?
Buying previously owned goods not only assists your budget but is also good for the environment! The article "9 Reasons to Shop Second-Hand" lists this and eight other positive aspects of shopping secondhand (Froelich 2013).

this chapter—which include how to make change, how to keep their money safe, and how to assign resale value—to better manage their money.

Students are also introduced to another form of income besides a wage or salary. Children are limited in their ability to hold a job, so a garage sale might be a source of income or a place to trade items. Also, money is not limitless, and good planning and decision making play a role in creating and managing a budget. Garage sales are a way to get a very good price for buying things you might want or need. One message to be taken away from garage sales is that whether or not something is sold, it does have value and must be treated with care. By beginning these discussions at a younger age, children will have more experience and perspective for when they become independent, primary consumers.

Student Knowledge and Experience

Students can struggle with mental computation and estimation, particularly if they are trying to employ a standard algorithm rather than use a more logical mental strategy. Explicit instruction on mental strategies may be needed. For example, a number line can be used to model how to use a jump-up or jump-down strategy for computing.

By school age, some students might already have experience with secondhand shopping. They may have accompanied an adult to a used car lot or thrift store or attended a swap or garage sale. Additionally, some students, particularly younger ones, may struggle with placing a monetary value on their possessions. They might price the antique doll that their grandma gave them for their birthday at a dollar, whereas the price for a favorite teddy bear could be a million! Students are likely to value their own possessions, and this can be a good way to enter into the discussion of how to put a price tag on something to sell to someone else.

Lesson Plan

Learning Targets

- Perform mental calculations in the context of shopping at a garage sale.
- Estimate purchases.
- Realize that paper money, coins, and other objects have value.

Resources and Tools

- **Activity Sheet 5.1:** Garage Sale (one for each student).
- **Resource 5.1:** Garage Sale Signs 1–3 (one set that is laminated for repeated use with future classes or multiple sets to be cut up and used by students in this class)
- *Yard Sale* by Eve Bunting (2015) (or another book about buying and selling previously owned goods)
- Various smaller items—pencils, erasers, and so on—that students can use in a mock garage sale (optional)
- Coins to spend (real or pretend pennies, nickels, and dimes; quarters depending on appropriateness)
- Optional: TI-Nspire™ activity: Garage Sale! (In this activity, students can create their own story about shopping at a garage sale, how much they plan to buy, how much money they have, how they calculate how much they will be spending, and how much they will

have remaining. Students drag each object into their cart, and the current total in cents is represented in their cart. Each student can create his or her own unique stories, with differing amounts of money.

- Optional: iPad® with TI-Nspire™ Garage Sale! app loaded.

Preparation Prior to Lesson

Part 1: Obtain a copy of *Yard Sale* by Eve Bunting. Before the lesson, set up a "garage sale" in the classroom. Items that you might use include pieces of candy, pencils, erasers, homework passes, small games, coloring books, and toy cars. You could also reproduce and use Resource 5.1: Garage Sale Signs for items that might commonly be found at a garage sale. Each item should have a monetary value and be clearly labeled for students to "purchase." Prepare enough bags of coins for each student in your class to have enough money to purchase items at your classroom garage sale. Make enough copies of Activity Sheet 5.1 for every student to have one.

Part 2: Gather items that students might want to buy, and create signs advertising the product with the cost for each item (optional).

Children's Literature Option

See the Literature for Financial Literacy list on p. 203 for additional relevant children's books.

Key Language

four basic mathematical operations: Addition, subtraction, multiplication, and division.

mental mathematics: Performing mathematical operations in your head without the aid of another mathematical calculation tool.

purchase: Something you buy with money.

value: The importance, worth, or usefulness of something, represented in money.

Money Talks

Begin the lesson with a literature connection, using *Yard Sale* by Eve Bunting, to get students thinking about yard or garage sales, their purpose, and what they might find to **purchase** at one. Talk with the students about the feelings the little girl was having in the story. Ask questions such as the following:

- Why do you think Callie's parents were having a yard sale?
- Why do you think Callie had a hard time selling her possessions?
- What were the benefits to Callie of selling some of her possessions?
- What do you think was the most important possession to Callie?

Through this book, students will learn that the most important thing we have is each other. Possessions are just that, things that can be sold for a price, but no one can put a price on the love within a family. This family had to have a yard sale to downsize and to make some extra money. Students will recognize that money doesn't always come from earned income at a job.

Sales Pitch

Talk with students about why people hold garage sales and what benefits might occur (e.g., making money, cleaning house, getting rid of items that are no longer used). Ask them the following questions:

- Is there anything in your bedroom or house that you no longer use and might want to sell?
- What **value** do you think these items have?

Make a list of these items on the board while talking about value, including the ways items can maintain their value.

Next, ask students to say what are some things that they might like to buy. Discuss how shopping at garage sales is sometimes a great way to find items they want and that—because they have been owned previously—might be purchased for a lower cost than something new. List the items that students name on the board. After collecting some ideas, assign cost values to the items (what you think they would cost at a garage sale in your area).

Using the created list, perform some **mental mathematics** calculations with the class using the **four basic mathematical operations**. Ask students questions such as these:

- If I bought three books at $1.00 each, how much did I spend?
- I am selling bookmarks for 4 for $2.00. How much would one bookmark cost?
- I bought a shirt for 50¢, a pair of pants for 50¢, and a pair of shoes for $1.75. How much did I spend altogether?
- What was the total that I spent?
- After I bought the above items, I decided to return the pants. How much did I spend now?

Make sure to verbalize your thought process and how you arrived at your answer. Next, model this mathematical thinking through equation representations.

The Fine Print

Part 1: Preparation

Hand students a copy of **Activity Sheet 5.1: Garage Sale** and a bag containing a collection of coins. Ask the students to take out their money and count the coins, recording the amount on Activity Sheet 5.1. Review the coins and their values, if necessary, as well as how to count a small collection of coins. All students should begin with the same amount of money and with denominations based on the appropriateness for the grade level. Explain to students that the money is not limitless and that this is all they have to spend at the sale, and this might require some financial decision making and planning on their part.

Instruct the students to leave their money on their desk and "window shop" only, by first browsing the items and how much they cost. This is a "look but don't touch" opportunity. Explain to students that they might have to do some mental mathematics to figure out if they have enough money to make their desired purchases. Once they go back to their seats, the students are welcome to make notes on their activity sheet, use a calculator, or perform paper-and-pencil calculations to aid in the planning process.

Part 2: Shopping

Explain to the students that they will first have the opportunity to make one purchase and then find another item for sale that they are going to put on their wish list. Students need to bring the item

Standards for Mathematical Practice: Students use MP.5 as they select whether to calculate mentally or to use a calculator, and they use MP.1 as they make sense of the situation and decide if they have enough money to make a second purchase.

that they would like to purchase, with their money, to the cashier. The cashier could be the teacher, a volunteer, or one student assigned to each small group. They will pay the cashier for the purchase and take their change back to their seat. On Activity Sheet 5.1, have the students record their first purchase price subtracted from the initial amount in the table to calculate their change. Before the students count their coins again, ask them to use mental math to calculate how much money they have remaining after making their first purchase. Students may verify their answer by counting the physical coins and then consult their wish list to determine if they have enough money to buy another item. Give students the option to purchase the item on their wish list or to save their money for a future shopping opportunity.

Closing the Deal

Once students have moved through the garage sale activity, come back together as a group and ask students to reflect on their strategies. Questions might include these:

- What strategies do we use when we want to add without writing it down on paper?
- What strategies do we use for subtraction?
- What strategies do we use for multiplication?
- What strategies do we use for division?

Review financial literacy concepts with questions such as these:

- Why might someone have a garage sale?
- When might someone shop at a garage sale?
- What are the pros and cons of selling and shopping for secondhand items?

Getting Your Money's Worth

This activity can be modified for multiple grade levels by using different collections of coins and amounts. Students can also have the chance to play different roles, such as the seller, consumer, or cashier. Students who excel at mental mathematics can scale the activity up to include multiple operations in the same calculation. For example, a student has 80 cents and wants to buy 2 pencils that cost 10 cents each and a bouncy ball for 50 cents. Does the student have enough money to buy the pencils and the ball, and if so, how much money would be left?

References and Resources

Bunting, Eve. *Yard Sale.* Somerville, Mass.: Candlewick Press, 2015.

Froelich, Amanda. "9 Reasons to Shop Second-Hand" (2013). http://www.trueactivist.com/9-reasons-to-shop-second-hand/.

Jump$tart Coalition for Personal Financial Literacy. *National Standards in K–12 Personal Finance Education.* Fourth Edition. Second Printing. Washington, D.C.: Jump$tart Coalition for Personal Financial Literacy, 2017.

National Day Calendar. "National Garage Sale Day" (n.d.). https://nationaldaycalendar.com/national-garage-sale-day-second-saturday-in-august/

National Governors Association Center for Best Practices and Council of Chief State School Officers (NGA Center and CCSSO). *Common Core State Standards for Mathematics.* Washington, D.C.: NGA Center and CCSSO, 2010.

Activity Sheet 5.1

Garage Sale

Record your starting amount of money in the table below. How many of each coin do you have, and what is their total value?

Coin Name	Value of One	How Many Coins?	Total Value
quarter	25¢		
dime	10¢		
nickel	5¢		
penny	1¢		
Add up the total value of all of your coins:			

Window Shopping Notes/Wish List:

Total Money from Above:	
Amount of First Purchase:	
Amount of Change:	

Do you have enough money to purchase an item from your wish list? Explain your answer in pictures or words.

Resource 5.1

Garage Sale Signs 1

Resource 5.1

Garage Sale Signs 2

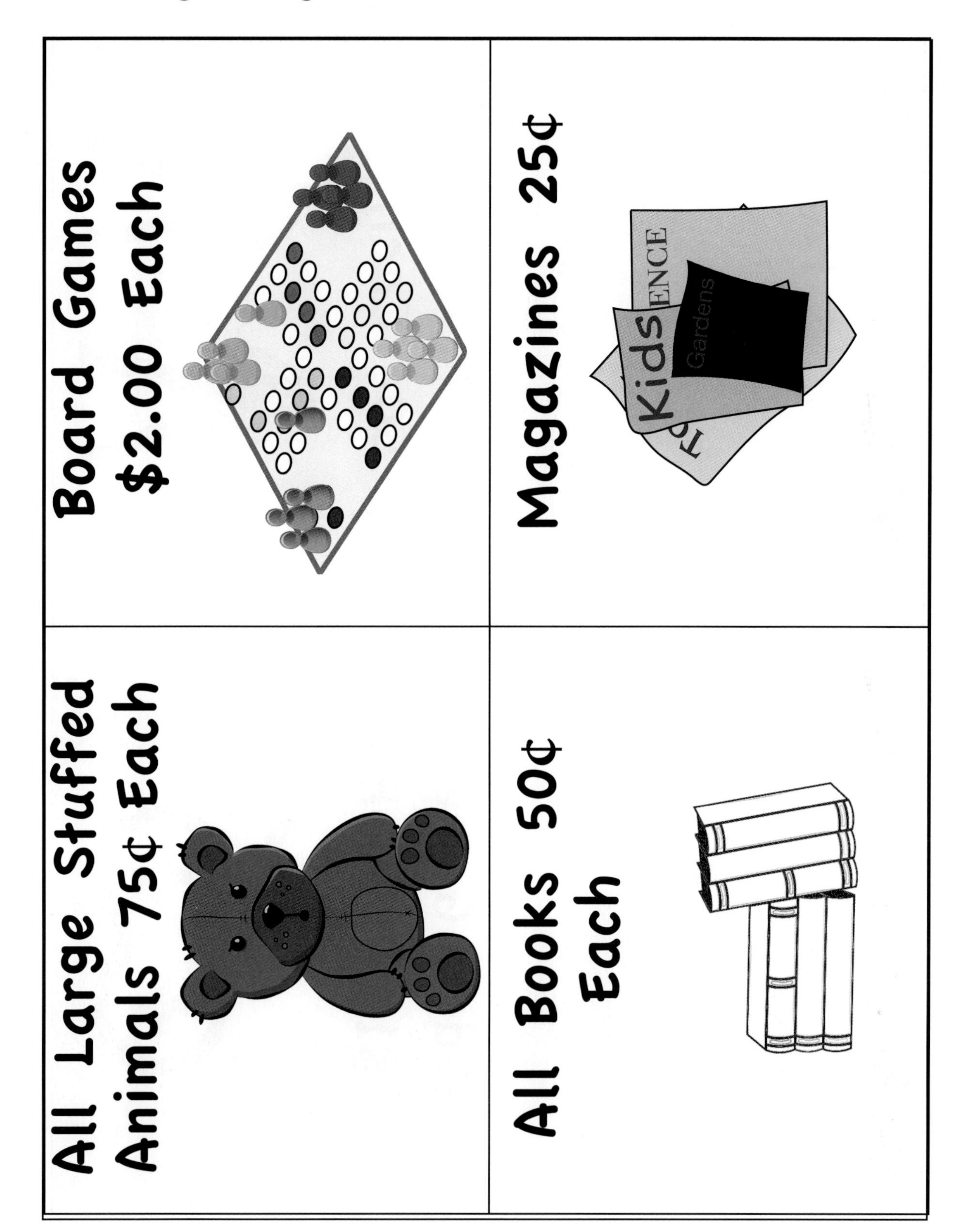

Resource 5.1

Garage Sale Signs 3

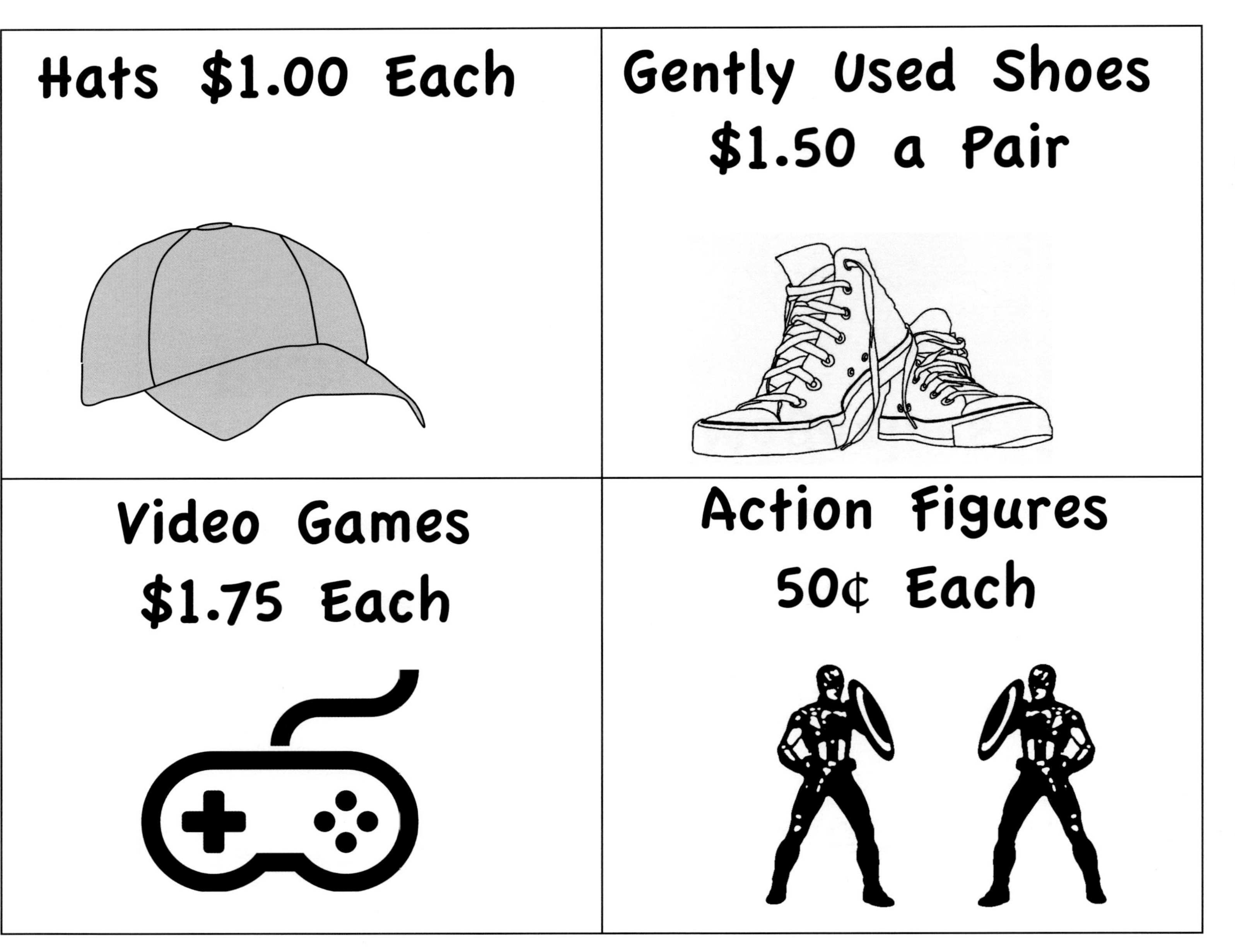

Chapter

Spend, Save, or Share?

The Bottom Line

Money is earned, and once earned, you have options of what to do with that money—spend it, save it, or share it.

Students solve problems involving the four basic mathematical operations and make financial decisions within the context of earning money.

Standards Met in This Chapter

CCSSM Standards and Practices

3.OA.D.8: Solve two-step word problems using the four operations. Represent these problems using equations with a letter standing for the unknown quantity. Assess the reasonableness of answers using mental computation and estimation strategies including rounding.

3.OA.D.9: Identify arithmetic patterns (including patterns in the addition table or multiplication table), and explain them using properties of operations. *For example, observe that 4 times a number is always even, and explain why 4 times a number can be decomposed into two equal addends.*

4.OA.A.3: Solve multistep word problems posed with whole numbers and having whole-number answers using the four operations, including problems in which remainders must be interpreted. Represent these problems using equations with a letter standing for the unknown quantity. Assess the reasonableness of answers using mental computation and estimation strategies including rounding.

4.OA.C.5: Generate a number or shape pattern that follows a given rule. Identify apparent features of the pattern that were not explicit in the rule itself. *For example, given the rule "Add 3" and the starting number 1, generate terms in the resulting sequence and observe that the terms appear to alternate between odd and even numbers. Explain informally why the numbers will continue to alternate in this way.*

5.OA.B.3: Generate two numerical patterns using two given rules. Identify apparent relationships between corresponding terms. Form ordered pairs consisting of corresponding terms from the two patterns, and graph the ordered pairs on a coordinate plane. *For example, given the rule "Add 3" and the starting number 0, and given the rule "Add 6" and the starting number 0, generate terms in the resulting sequences, and observe that the terms in one sequence are twice the corresponding terms in the other sequence. Explain informally why this is so.*

MP.1: Make sense of problems and persevere when solving them.

MP.2: Reason abstractly and quantitatively.

MP.4: Model with mathematics.

(National Governors Association Center for Best Practices and Council of Chief State School Officers [NGA Center and CCSSO] 2010)

Jump$tart Standards

Spending and Saving: Apply strategies to monitor income and expenses, plan for spending and save for future goals.

Knowledge Statements:

Overall:

- Saving means choosing not to spend money today in order to buy something in the future.

Financial Decision Making: Apply reliable information and systematic decision making to personal finance decisions.

Standard 1. Recognize the responsibilities associated with personal financial decisions.

Knowledge Statements:

Standard 1:

- Predict the consequences of spending decisions.

(Jump$tart Coalition for Personal Financial Literacy 2017)

Future Value

Giving students choices and helping them make decisions not only aids in the development of literacy but also supports good decision-making strategies in general. Consequently, students witness firsthand how mathematics is used to make important decisions. Learning to consider consequences when making choices helps students become thoughtful decision makers, whether or not they are making decisions that involve money.

In this chapter, we discuss the importance of balancing money that is spent with money that is saved and shared. Learning to make these financial decisions now is a precursor to future financial literacy concepts such as budgeting and investing. Keeping an appropriate balance among the three helps create positive money-handling habits that students will be able to carry over into the future.

Mathematics

This lesson focuses on developing students' meaning of multiplication and division of whole numbers through performing calculations in problem-solving situations involving spending (3.OA.D.8; 4.OA.A.3; MP.1), sharing, and saving money. Students will also be exposed to recognizing patterns and finding rules for calculating how much will be saved, spent, and shared based on an hourly wage (3.OA.D.9; 4.OA.C.5; 5.OA.B.3). Students will need to reason abstractly and quantitatively when making an educated guess as to how long it will take them to save for their intended purchase, and they will apply mathematical knowledge when creating a model to represent the situation (MP.2; MP.4). This chapter also can be connected to place-value concepts, rounding, estimation, and decimals, depending on the grade-level focus.

Financial Literacy

When you make money, you have three options of what to do with it: spend it, save it, or share it. Each of these options comes with trade-offs. For example, if you spend your money, you do not have it for something you want or need later, and you do not have the option to share it with others. Financial literacy includes developing an understanding and appreciation of why a person should save money and also how sharing money can benefit other people or support the community or other causes. When students have a conceptual understanding of sharing, spending, and saving their money, they are building a foundation from which they can later explore investing and budgeting.

Did you know? ThreeJars.com is an excellent site for exploring the concept of spend, save, share. The philosophy is based on children taking on the responsibility to learn positive money habits (Threejars.com n.d.).

Student Knowledge and Experience

Students are likely to have had previous experience with earning money; familiarities with the ideas of spend, save, and share; and they may be aware of the benefits of each choice. Even if they have not had the opportunity to donate money through sharing, connections can be made to the time they have done something nice for another person. For example, maybe they shared their time by volunteering at a soup kitchen during the holidays or they helped rake a neighbor's leaves. Or maybe they let another student borrow a pencil or shared their art supplies during a school project.

Mathematically, the notion of sharing can occur with the four mathematical operations (most frequently, division). Though students have had experience with these operations, they may not be as comfortable with estimation or using mental strategies for computation. This chapter will help model these concepts to support student reasoning and encourage sense making.

Lesson Plan

Learning Targets

- Perform basic mathematical operations within the context of spending, saving, and sharing money.
- Describe options for what to do with earned money, and the advantages and disadvantages of the options.

Resources and Tools

- **Activity Sheet 6.1:** Spend, Save, Share (one for each student or pair of students)
- **Resource 6.1:** Money Uses Cards (one set for each small group of students)
- *Money Mama & The Three Little Pigs* by Lori Mackey (2003) (or another book about spending, saving, and sharing)
- Calculators (optional)

Preparation Prior to Lesson

Sales Pitch: Create card decks out of the Resource 6.1: Money Uses Cards.

Part 1: Obtain a copy of *Money Mama & The Three Little Pigs* by Lori Mackey.

Part 2: Prepare enough copies of Activity Sheet 6.1 for each student. Calculators are optional.

Children's Literature Option

See the Literature for Financial Literacy list on p. 203 for additional relevant children's books.

Key Language

earn: To gain money, sometimes in exchange for work.

need: Something that is required to live.

save: When you keep your money somewhere safe to be used to spend on something later.

share: When you give your money to other people who need it.

spend: When you use money to get something you want or need.

want: Something that you do not need, but you desire to have.

Money Talks

This is a two-part "Money Talks." First, ask students how they might **earn** or receive money. This is a great opportunity to discuss with students that they might get money from doing work but also might get some from other sources, such as a birthday gift.

Second, focus on what they might do with money that they have. If you did the Garage Sale! lesson in chapter 5 and students earned money from it, connect to those specific examples. Otherwise, pose the following question to students:

- You earned $25 selling toys and clothes at a garage sale. Design a plan for what you will do with your earnings and be ready to share!

Invite a collection of five to ten responses. Discuss patterns across these answers (which can be mathematical patterns, as well as financial patterns).

Sales Pitch

Place students with partners or small groups. Give students a stack of the cards from **Resource 6.1: Money Uses Cards.** Ask them to sort them into categories of how the money was used. After each group has sorted its cards, take time to compare how they were sorted. One way to do this is to have groups share. Another way is to hold up a particular card and ask different groups what group label they created for that money use. In the end, the hope is for students to come up with categories of **spend, save,** and **share.**

The Fine Print

Part 1: Wants versus Needs

Read the book *Money Mama & The Three Little Pigs* by Lori Mackey to get students thinking about the concepts of spending, saving, and sharing. Have students take out their financial portfolio or mathematics journal and list one way that they have (1) spent money, (2) saved money, and (3) shared money. Give a few students the opportunity to share with the large group.

Ask students to describe the difference between buying something you want and buying something you need. Students might have trouble making this distinction. Create a list of students' **wants** and **needs.** Discuss with students whether they would categorize each item in the list as a want or a need. After the two lists are complete, ask students to describe the differences between the items on the list of wants versus items on the list of needs. Have students reflect on something they needed that either they bought themselves or that their family purchased.

Part 2: Spending Money

Begin by having students think about something they, or someone close to them, have purchased in a store recently. If they were the one to make the purchase, have students reflect on how much they spent on the item. Ask the students the following questions:

- Was this an item you still use frequently?
- Was this a planned purchase or something that was bought on impulse?
- How did you get the money to make this purchase?

Have students also think about some items that they borrowed and didn't need to purchase—such as borrowing a book from a library as opposed to buying the book from a bookstore. Ask students the following:

- Could you have borrowed the previously purchased item?
- Would you still choose to spend your money on this purchase?

Standards for Mathematical Practice: Students develop MP.1 as they use the mathematics to make sense of the problem and figure out how long it will take to save the money for their purchase.

Next, have the students think of two items that they really want to buy. Ask the students the following questions:

- What are two things that you would want to purchase?
- How much money do those two items cost?
- How are you going to be able to pay for the two items you want?

Have students complete **Activity Sheet 6.1: Spend, Save, Share** as a hypothetical exercise in spending, saving, and sharing. Discuss with students the patterns that they recognized when completing this activity. As a class, create a rule for the chart on the second page of the activity sheet. Ask students the following:

- We have fractions to represent the amount of our wages that goes in each jar, but what operation are we using when separating the money?
- What pattern do you see in the save column? The share column? The spend column?
- Which column is growing faster? Why?
- What rule would apply to each column?

Make sure to discuss the last question and the benefits of using the spend, save, share model. Though the process may take longer when saving the money for the items, remind them that they are also helping others *and* saving for the future!

Closing the Deal

Once students have moved through the Spend, Save, Share activity sheet, ask them to think about things they learned, things they wonder, and things they had trouble understanding. Ask the following:

- What patterns did you notice when completing the chart?
- Why do you think that happened?
- Why do you think it is hard for people to put their money in the save and share categories?
- What are the benefits of doing so?

Suggest that students go home and find three small jars, boxes, or envelopes, and label them "spend," "save," and "share." As students earn money, encourage them to separate it into categories as a way to develop their decision-making and budgeting skills. Students can take what they have learned in class and apply it to everyday life!

Getting Your Money's Worth

This activity can be modified for multiple grade levels by asking students to think ahead about some things they will need or want in the future and to consider how they are going to pay for them. In particular, what are big items that require saving over a long period of time (cars, college, video game systems, etc.).

References and Resources

Jump$tart Coalition for Personal Financial Literacy. *National Standards in K–12 Personal Finance Education.* Fourth Edition. Second Printing. Washington, D.C.: Jump$tart Coalition for Personal Financial Literacy, 2017.

Mackey, Lori. Money *Mama & The Three Little Pigs*. Agoura Hills, Calif.: P4K Publishing, 2003.

National Governors Association Center for Best Practices and Council of Chief State School Officers (NGA Center and CCSSO). *Common Core State Standards for Mathematics*. Washington, D.C.: NGA Center and CCSSO, 2010.

ThreeJars.com: Save, Spend & Share [website]. http://www.threejars.com.

Activity Sheet 6.1

Spend, Save, Share

This is a hypothetical exercise in spending, saving, and sharing money. You will not actually be able to purchase your wanted items, but planning for the future might help you save for items you want to buy later.

Name of the item you would like to purchase	Cost of the item
1.	
2.	

What are some ways that you could *earn* money to purchase your items?

Congratulations! You got a job delivering newspapers in your neighborhood and will be making $10 an hour. Answer the following questions based on your hourly wage. Be sure to show your work!

How many hours would you have to work to purchase item #1?

How many hours would you have to work to purchase item #2?

How many hours would you have to work to purchase both items?

Activity Sheet 6.1—*page 2*

Spend, Save, Share

In school you have been learning about the importance of saving and sharing your money. You have decided to put your earnings into three jars. 2/5 of your money goes into saving and 1/5 of your money goes into sharing. That leaves 2/5 of your money to spend. How much of your hourly wage is available for spending? Use the chart below to help you organize your calculations.

Number of Hours	Save	Share	Spend

Using this new information, answer these questions based on your hourly wage. Be sure to show your work!

How many hours would you have to work to purchase item #1?

How many hours would you have to work to purchase item #2?

How many hours would you have to work to purchase both items?

How much longer would it take you to purchase the items now than it did before?

Why might it be a good idea to use the save, spend, share model rather than purchasing something impulsively?

Resource 6.1

Money Uses Cards

Went to the movies.	Pledged money for a walk-a-thon.	Bought shoes.
Put money in my bank account.	Paid for a snack at lunch.	Bought tickets to the zoo.
Put coins in my piggy bank.	Bought socks for a sock drive at school.	Rented a bike for a day.
Hid money in my room.	Paid dues for a club or sport.	Put coins in my sister's piggy bank.
Bought a gift card for a friend.	Paid to download a song.	Bought a book.
Gave money to my cousin.	Bought a game at the store.	Gave money to parent to save for me for later.

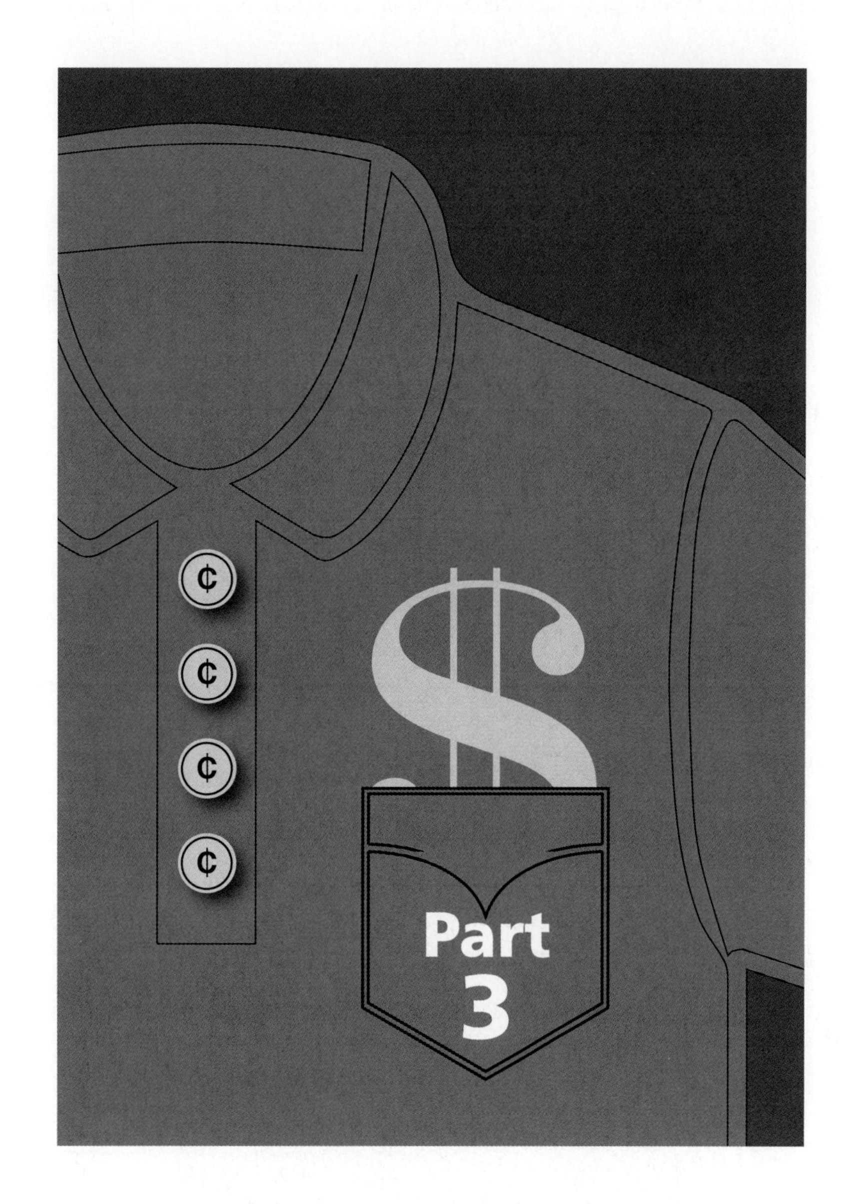

Number and Operations in Base Ten

Chapter

Saving for the Future

The Bottom Line

Students will use place value to explore the financial concept of saving for the future.

Through two activities and a game, students use pennies, dimes, and dollars to explore place-value concepts in the context of saving money. Students reflect on how much time it takes to save money, recognizing that the more money that is saved, the faster a savings goal can be reached.

Standards Met in This Chapter

CCSSM Standards and Practices

1.NBT.B.2: Understand that the two digits of a two-digit number represent amounts of tens and ones.

1.NBT.C.5: Given a two-digit number, mentally find 10 more or 10 less than the number, without having to count; explain the reasoning used.

2.NBT.A.1: Understand that the three digits of a three-digit number represent amounts of hundreds, tens, and ones; e.g., 706 equals 7 hundreds, 0 tens, and 6 ones. Understand the following as special cases: a 100 can be thought of as a bundle of ten tens—called a "hundred."

2.NBT.B.5: Fluently add and subtract within 100 using strategies based on place value, properties of operations, and/or the relationship between addition and subtraction.

MP.5: Use appropriate tools strategically

MP.7: Look for and make use of structure

(National Governors Association Center for Best Practices and Council of Chief State School Officers [NGA Center and CCSSO] 2010)

Jump$tart Standards

Spending and Saving: Develop a plan for spending and saving.

Standard 1: Apply strategies to monitor income and expenses, plan for spending and save for future goals.

Standard 2. Develop a system for keeping and using financial records.

Knowledge Statements:

Overall:

- Saving means choosing not to spend money today in order to buy something in the future.
- Saving means choosing to set aside money for emergencies and future needs and goals.

Standard 1:

- Share an experience of waiting to have enough money to buy something.
- Tell about a personal savings goal in terms of a special occasion in the near future, such as a gift or special event.

Standard 2:

- Create a way to keep track of money saved for future spending.

(Jump$tart Coalition for Personal Financial Literacy 2017)

Future Value

Learning the importance of saving money, including how long it might take to save up for a big item, has significant future value. Saving money can help avoid going into debt and can help in purchasing big items such as a car, a college education, or a home. Mathematically, developing place-value concepts also has significant implications for future mathematical topics. Gojak and Miles (2015) write, "Place value is the foundation for all future work with whole numbers and decimal numbers. Through the use of concrete and pictorial models and later connecting those models to symbolic notation is fundamental to developing conceptual understanding" (p. 77).

The following sections describe the mathematical and financial literacy concepts that this lesson will encompass, as well as the knowledge and experiences that elementary students may bring to the lesson and take from it on its completion.

Mathematics

Place-value understanding requires the physical grouping of ten ones to create one ten and then engaging in counting these groups of ten along with groups of one (1.NBT.B.2). In part 1, students physically group ones in little cups to see how much money they have saved. After sufficient experiences with groupable models, students can transition to representational models. Money is such a model. The dime is not physically 10 pennies; it just represents 10 pennies. As students explore the physical grouping of pennies, they are developing an understanding of the structure of numbers (MP.7).

In part 2 of the lesson, students keep track of loose change (dimes and pennies) that have been found to see how much money has been saved. In part 3, students continue to use a representational model—this time, one-dollar and ten-dollar bills—to determine how long it might take to save a certain amount of money. This final activity is in a game format (much like Race to 100). After students have achieved a deeper understanding of place value, they are able to begin mentally adding and subtracting with 10 more or 10 less and progress into adding multiples of 10 (1.NBT.C.5; 2.NBT.A.1; 2.NBT.B.5). In these latter two parts, a bottom-up hundreds chart (Counts "up" to 100, rather than "down" to 100) (Bay-Williams and Fletcher, 2017) and number line are used to support student thinking about adding 10s and adding 1s (MP.5).

Financial Literacy

Did you know? Recall the old adage: "If you can't afford to pay for it now, you can't afford it"? According to the website CreditLoan.com, the average American will pay more than $600,000 in interest over the course of a lifetime! Yikes!

According to Whitebread and Bingham (2013), "By the age of seven years, several basic concepts relating broadly to later 'finance' behaviours will typically have developed" (p. 17). Among these is an understanding that money is used to purchase things. Seeing that money is exchanged for goods is just the beginning of financial literacy. In order to trade money for goods, one might possess the money in the first place! While young students may have real-world encounters where they witness spending, it is less likely that they have witnessed saving. A central aspect of financial literacy is having a spending plan and a saving plan. Saving has certainly been underemphasized in recent years. When we do not save money to buy what we need, we have to borrow money. This makes the cost of the items more than the price tag!

In this chapter, the students verbalize their own experiences with saving. They discuss what it feels like to wait for something and what planning and decisions are involved in making a big purchase. Through a series of activities and games, students think about how long it might take them to save for something they might want or need.

Student Knowledge and Experience

Through their everyday observations and activities, students will have various experiences with spending and saving. Some will struggle with giving up their money to make a purchase. Others will have difficulty with the concept of waiting and saving for a future need or want. Regardless of their starting point, remember that "forward" thinking is very abstract to young children. By linking the future to recognized events—such as days of the week, holidays, or birthdays—students will start to make connections to concrete occurrences, and thus the concept becomes more easily understood.

Lesson Plan

Learning Targets

- Use tools and models to add tens and ones within 100.
- Use place value to explore the financial concept of saving for the future.

Resources and Tools

- **Activity Sheet 7.1:** Loose Change Adds Up!
- **Resource 7.1:** Bottom-Up Hundreds Chart (one for each student or pair of students)
- **Resource 7.2:** Number Line template, 0–100
- Bag of pennies (100 pennies for each student or pair of students)
- Small paper cups (10 for each student or pair of students)
- Bag of 10 dimes (plastic or real) per group
- 2 dice per group

Preparation Prior to Lesson

Part 1: Prepare enough copies of Resource 7.1 as needed for each student. Laminate a class set for reuse. Prepare a bag of 100 pennies for each student or pair of students.

Part 2: Prepare a bag of 10 dimes for each student or pair of students. Copy Activity Sheet 7.1 as needed for each student.

Part 3: Prepare enough copies of Resource 7.2.

Children's Literature Option

See the Literature for Financial Literacy list on p. 203 for ideas of relevant children's books.

Key Language

place value: How much a digit is worth based on its location in a number.

savings: The amount of money not spent and kept in a bank or other safe place for a period of time.

Money Talks

Ask your students to think of a time when they had to wait a long time for something (provide an example). Ask students to share what they had to wait for, and record ideas on the board. Follow up with questions about what they waited for: How long did they have to wait? What might be reasons we have to wait for something? Talk with students about the notion of time and how when we are waiting we could count down the number of months, days, or hours. For instance, you could discuss how many "sleeps" it is until the weekend, or how many calendar pages it is until their birthday. Explain to students that the time you wait for something that costs money depends on how much time it takes to save up that money.

Sales Pitch

Ask students if they have ever found coins on the ground. What did they do? (pick them up, leave them there). Ask students if they have heard of these quotes (select some or use all):

"See a penny, pick it up, all day long, you'll have good luck!"

"A penny saved is a penny earned." [It is as useful to save money that you already have as it is to earn more.]

"Waste not, want not." [If you don't waste food/money, then you won't end up without it.]

Discuss what they mean. Ask students to share their own experience with saving money (in the past or present). They may do this in writing, by talking to a partner, or in a whole group.

Explain to the students that today they will be exploring the idea of saving money and seeing how long it takes to save an amount of money they need.

The Fine Print

Part 1: Piles of Pennies

To begin, ask students to count how many pennies they have in their bag. Give the students the opportunity to come up with their own grouping strategy. Look to see if students are making stacks or groups of 10 as they count. If not, discuss how coins can be grouped in order to count by groups.

Standards for Mathematical Practice: As students use the hundreds chart or number line to quantify the pennies, they are strategically using a mathematical tool (MP.5) and making use of patterns and structure (MP.7).

Ask students to now group their pennies in cups of 10 cents. That is, to place ten pennies in each of their cups. Once their cups are full, practice skip-counting by tens with students. Use a hundreds chart or number line to add another model to support student thinking.

Next, distribute a copy of **Resource 7.1: Bottom-Up Hundreds Chart** to each group. Explain that students will be working with a partner. On their turn, they point at (or cover) a number on the hundreds chart. Their partner places that quantity of pennies on the chart using the cups of coins. For example, if a student points at 30, his or her partner places three cups on the chart. If a student points at 52, then his or her partner places 5 cups and pulls 2 pennies out of a different cup and places them on the mat. In addition or alternatively, students can use a 0–100 number line, placing a cup at each 10, and the pennies on each one.

Part 2: Loose Change Adds Up!

Talk with the students about why we might count by tens rather than always choose to count by ones. (Counting by ten is faster and more efficient.) Next, mention to the students that there is another coin that also has a value of 10 cents. Show the students a dime and give them the opportunity to explore its attributes. Teach the students the rhyme, "Dime, dime, little and thin, I remember you're worth ten." To give students a visual of why we might choose to count by ten, demonstrate a baggie filled with 100 pennies and a baggie filled with 10 dimes. Explain to the students that both bags are equal in the value of their money, even though the penny bag has ten times as many coins. Ask them which might be more comfortable to carry if they had to walk around with 100 cents in their pocket. Have the students trade their stacks of ten pennies for dimes. [Note: Part 1 activities can be repeated using dimes and pennies.]

Standards for Mathematical Practices: Students exhibit MP.5 when they consider the available tools and select an option that can help them add by tens and by ones.

Saving to buy something can start small. In this activity (see **Activity Sheet 7.1: Loose Change Adds Up!**), Daniel wants to save coins in order to have a dollar he can spend at the dollar store. Distribute pennies and dimes for students to use. Invite students to select a tool that can help them keep track of how much money they have saved—a hundreds chart (see Resource 7.1) or a number line (see **Resource 7.2: Number Line**). Additionally, students may benefit from using base-10 pieces or other **place value** tools. Discuss and model, as needed, how to use these tools to keep track of how much money is found and how much money is saved. For example, you may want to have students use their selected tool to determine how much money was found and saved on Monday and then stop to have students share their thinking before moving to the other days of the week.

After students have completed Activity Sheet 7.1, ask the following questions:

- What patterns did you notice when you added a dime to your savings?
- What patterns did you notice when you added a penny to your savings?
- Did it take more or less time than you thought to save loose change to equal $1?
- If Daniel kept looking for loose change, how long might it take him to find $10? $20?
- Imagine that Daniel bought some gum during the week. Would he have had enough money by the end of the week?
- Why might we save money? What can be difficult about saving money?

Part 3: Saving for a Big Item

Show an advertisement of an item that costs $80 to $100 (or a higher value for students who are ready). Ask, "If I start saving money today, how long will it take before I can buy [this item]?" Listen to different ideas. Of course, the answer depends on how much you save! A student might explain that you could save $10 every day and buy the item in 8 days (or every week, and then buy in 8 weeks). Skip-count on the calendar or hundreds chart to explore the ideas here. Explain to students that sometimes we have to wait to buy something we really want, so we save money in order to buy something in the future.

To explore how long it takes to save $100, play Save to Your Goal (a modification of the Race to 100 Game), as explained in figure 7.1. Invite students to select the hundreds chart (Resource 7.1) or number line (Resource 7.2) to keep track of their **savings.** In the end, they must share how many days it took them to save $100.

Save to Your Goal of $100

Goal: Save $100 in fewer days than it takes your opponent.

Materials: Two dice, hundreds chart, or number line (optional)

How to Play:

1. On a player's turn, he or she rolls two dice. That sum represents how much money the player earns and saves on that day.
2. Each player records his or her savings in a table, recording how many days and the total savings.

Day (Roll)								
Savings								

3. Play goes to the next player.

Optional Rules:

4. If a player rolls two 1s ("snake eyes"), he or she didn't save any money (lose a turn).
5. If a player roll two 2s ("twos blues"), he or she spent $10 (go down 10).

Fig. 7.1. Rules to "Save to Your Goal"

Save to Your Goal can be played using pennies and dimes, with the focus on saving $1.00. After students have played a few rounds of the game, discuss the mathematics and the financial literacy concepts. Ask the following questions:

- Which tools or models did you use to keep track of your money?
- What strategies did you use to add or subtract 10? To add ones?
- What was a common number of days it took to save $100?
- What kinds of things affect how long it takes to save $100?
- In a real situation where you wanted to save $100, how might you earn that money?
- How long might it take to save that much money?
- How might you keep track of how much money you have?
- Why might you want to save money even if you don't have something specific you want to buy?

Closing the Deal

Close the lesson by asking students to think about how adults save money, where they save their money, and how they might know how much money they have saved. Help students to develop their own savings plan. For each question, collect several ideas from students and then have them write their own choice in a journal:

* What might you want to save money? (For an item to buy, for an emergency, to do an activity, etc.)
- What amount of money is your personal goal?
- How might you get the money?
- Where will you save the money?
- How will you keep track of your money?

Getting Your Money's Worth

This activity is great for multileveled students within one classroom or for various grades. Once students have mastered pennies and dimes, they can start to incorporate dollars as the hundreds place. Or, they can use bills for $1, $10, and $100.

Schools are regularly doing fundraisers—in effect the school PTA is trying to earn and save money for a particular need for the school. Students can investigate what their school is saving for, what the item costs, and how much time it might take to save up for that particular item.

References and Resources

Bay-Williams, Jennifer M., and Graham Fletcher. "A Bottom-Up Hundred Chart?" *Teaching Children Mathematics* 24, no. 3 (2017): 153–60.

Gojak, Linda M., and Ruth E. Harbin Miles. *The Common Core Mathematics Companion: The Standards Decoded, Grades K–2: What They Say, What They Mean, How to Teach Them* (Corwin Mathematics Series). Thousand Oaks, Calif.: Corwin, 2015.

Jump$tart Coalition for Personal Financial Literacy. *National Standards in K–12 Personal Finance Education*. Fourth Edition. Second Printing. Washington, D.C.: Jump$tart Coalition for Personal Financial Literacy, 2017.

National Governors Association Center for Best Practices and Council of Chief State School Officers (NGA Center and CCSSO). *Common Core State Standards for Mathematics*. Washington, D.C.: NGA Center and CCSSO, 2010.

Whitebread, David, and Sue Bingham. *Habit Formation and Learning in Young Children*. London: Money Advice Service, 2013.

Activity Sheet 7.1

Loose Change Adds Up!

Daniel wants to go to the dollar store with his friends and pick out a toy. His mom says, "Why don't you take some time each day to look around the house for loose change? It might be in the couch, under the bed, and other places." Daniel agrees. Keep track of the money he finds each day, and figure out if he has made it to his goal of $1.00.

	Monday	Tuesday	Wednesday	Thursday	Friday	Saturday
Amount of money found						
Total money saved						

DAY	Loose Change
Monday	Daniel finds 2 dimes and 3 pennies.
Tuesday	Daniel finds another dime and 4 more pennies.
Wednesday	Daniel finds 2 more dimes. He saves them both.
Thursday	Daniel finds a dime and 3 pennies. He adds those to his savings.
Friday	Daniel finds 7 pennies and places them with his savings.
Saturday	Daniel finds 3 dimes and 1 more penny.

1. Did Daniel reach his goal?

2. How long might it take Daniel to save $10?

3. What things influence how long it takes to save money?

Resource 7.1

Bottom-Up Hundreds Chart

91	92	93	94	95	96	97	98	99	100
81	82	83	84	85	86	87	88	89	90
71	72	73	74	75	76	77	78	79	80
61	62	63	64	65	66	67	68	69	70
51	52	53	54	55	56	57	58	59	60
41	42	43	44	45	46	47	48	49	50
31	32	33	34	35	36	37	38	39	40
21	22	23	24	25	26	27	28	29	30
11	12	13	14	15	16	17	18	19	20
1	2	3	4	5	6	7	8	9	10

Resource 7.2

Number Line *Cut out each line, and tape them end to end.*

Buy Now and Pay Back Later

The Bottom Line

What is the cost of borrowing money?

Students mentally solve various scenarios of borrowing money (e.g., $30 or $400) and paying it back without a fee, and then the scenario of paying a fee of $1 for every $10 borrowed.

Standards Met in This Chapter

CCSSM Standards and Practices

1.NBT.B.2.c.: The numbers 10, 20, 30, 40, 50, 60, 70, 80, 90 refer to one, two, three, four, five, six, seven, eight, or nine tens (and 0 ones).

1.NBT.C.5: Given a two-digit number, mentally find 10 more or 10 less than the number, without having to count; explain the reasoning used.

1.NBT.C.6: Subtract multiples of 10 in the range 10-90 from multiples of 10 in the range 10-90 (positive or zero differences), using concrete models or drawings and strategies based on place value, properties of operations, and/or the relationship between addition and subtraction; relate the strategy to a written method and explain the reasoning used.

2.NBT.B.8: Mentally add 10 or 100 to a given number 100–900, and mentally subtract 10 or 100 from a given number 100–900.

MP.7: Look for and make use of structure.

MP.8: Look for patterns in repeated reasoning.

(National Governors Association Center for Best Practices and Council of Chief State School Officers [NGA Center and CCSSO] 2010)

Jump$tart Standards

Credit and Debt: Develop strategies to control and manage credit and debt.

Standard 3. Apply strategies to avoid or correct debt management problems.
Knowledge Statements:
Overall:

- Sometimes people can borrow money or an item if they promise to return it.
- Borrowing money to buy something usually costs more than paying cash because there is a price (interest and fees) for buying on credit.
- Borrowers who repay as promised show that they are worthy of using credit in the future.

Standard 3:

- Relate overspending to potential borrowing problems.

(Jump$tart Coalition for Personal Financial Literacy 2017)

Future Value

The cost of borrowing money is one of the most important financial literacy lessons for all of us. While elementary-age students are not yet ready to explore credit cards or short-term and long-term borrowing options, they are very ready to explore borrowing from a sibling or other relative.

In these activities, students are borrowing money (in multiples of 10 or 100), determining the cost of the loan ($1 per every $10 they borrow), and keeping track of their money balance. This chapter shares two parallel lessons involving mentally adding multiples of 10s (using numbers within 100) or multiples of 100s (using numbers within 1000). Therefore, the first lesson might be used in first grade and the second in second grade, or they both may be used in second grade, with the first serving as a scaffold (or a way to differentiate the lesson, giving the teacher or the students a choice of which one to explore).

The following sections describe the mathematical and financial literacy concepts that this lesson will include, as well as the knowledge and experiences that elementary students may bring to the lesson and take from it.

Mathematics

Mentally adding 10s or multiples of 10s (or 100s and multiples of 100s) is central to this lesson, as well as to developing procedural fluency. When students can mentally add or subtract a 10 or 100, they are able to use that strategy for adding or subtracting any number (1.NBT.C.5; 2.NBT.B.8). This is what is meant by "using place value strategies." Once students can add or subtract by 10 or 100, they can start chunking and mentally add or subtract a multiple, such as 30 or 400, which may initially involve skip-counting by 10s or 100s, but will eventually become adding 3 to the tens place (or 4 to the hundreds place) as students are asked to look for patterns in their work (1.NBT.B.2.c; 1.NBT.C.6; MP.8).

One of the most important ideas in the elementary curriculum is understanding that it takes 10 ones to make a ten, and conversely that a 10 is composed of 10 ones (MP.7). Because these loans are set up as $1 for every $10, the cost of borrowing $100 will be $10 and so on. While these young students are not exploring multiplication and division explicitly, they are informally seeing the place-value relationship between 1s, 10s, and 100s (that it takes 10 ones to make a 10, and 10 tens to make 100). This concept is embedded in this lesson as students think, "I need to pay $1 for each $10." For example, when reasoning about the cost of borrowing $30, they must see $30 as composed of 3 10s. Each 10 costs $1, so the cost is $3. With several examples, students will see that a $30 loan costs $3, a $40 loan costs $4, and so on (MP.8).

Financial Literacy

Did you know?
About *one-third* of the average American's take home pay is spent on interest (the fee for what they borrowed—this doesn't include what they spend paying back the money they borrowed) (Robbins 2014).

Borrowing begins even before children enter school (e.g., borrowing a friend's toy or checking out a book from the library). They may have even borrowed money and had to pay it back, for example, to a sibling or parent. But borrowing money is a big and important part of a person's financial well-being. Sharing more examples of when people borrow money is important in building students' understanding of debt. Examples include using a credit card or taking out a loan (such as from a bank, payday loan business, or pawn shop). In all of these cases, the person borrowing has to pay back the money and he or she have to pay a little more back than what he or she borrowed. Borrowing money, then, typically costs more than paying cash because there is a price (interest and fees) for buying on credit. This can become a problem if you borrow money and then can't pay it back—that is called *overspending.* Borrowing should not be positioned as something that is bad; the financial literacy lesson is that people must decide if the cost of borrowing the money is "worth it" for what they need. (Think of borrowing money for a car, house, or college.)

Student Knowledge and Experience

This lesson builds on and supports students' emerging understanding of place-value concepts and mental computation. Some students will be able to mentally add 42 + 30, already understanding they are adding 4 tens and 3 more tens. Others will benefit from tools to support their thinking. The most concrete tools are groupable models, like Unifix cubes (in stacks of 10s and loose). Then, other concrete tools like base-10 blocks. Next would be representational tools:

- A "bottom-up" hundreds chart, which is an excellent tool to support counting up and back by tens. (Note that when the chart counts up, the language of going "up" and "increasing" are aligned).
- A number line where the 10s (or 100s) are labeled, and there are smaller tick marks for the ones, encouraging the skip-counting by 10s or 100s.

Mental computation is valued highly in many countries (whereas the United States tends to focus more on "showing steps"). Students from other countries therefore may excel at these mental activities and be able to share their thinking with others. How students think about the numbers in their heads is likely to vary. Inviting students to share their thinking can help each child develop more efficient ways to add or subtract mentally.

Lesson Plan

Learning Targets

- Add and subtract 10 to or from a number between 0 and 100 (or 100 between 0 and 1000)
- Determine how many tens are in a multiple of ten (and number of 10s in multiples of 100s)
- Add and subtract multiples of 10 between 0 and 100 (or multiples of 100 between 0 and 1000)
- Describe reasons and related risks for borrowing money

Resources and Tools

- **Activity Sheet 8.1:** Lend and Borrow (one for each student)
- **Resource 8.1:** Ten Frame Cards
- **Resource 8.2:** Bottom-Up Hundreds Chart (one for each student)
- **Resource 8.3:** Number Line template, 0–100
- Bags to hold sets of Ten Frame Cards (one for each pair of students)
- One die
- *Brody Borrows Money* by Lisa Bullard (2013) (or another book about borrowing money)

Preparation Prior to Lesson

Part 1: Obtain: *Brody Borrows Money* by Lisa Bullard. In plastic baggies, place 12 full ten frame cards and two of each partially full ten frame card. One baggie is needed per pair of students. Full ten frame cards represent $10 bills, and partially full frames represent the number of $1 bills.

Children's Literature Option

See the Literature for Financial Literacy list on p. 203 for additional ideas of books to incorporate into the lesson.

Key Language

borrow: Take something, such as money, temporarily and then return it.

fee: The extra money you pay to the business or person who lent you money.

loan: Money you borrow and must pay back.

Money Talks

Ask children to think of something they have **borrowed** from someone. Begin with your own example (for example, you might say, "This morning, I had to borrow these cards from Ms. Eaton next door"). Give students a few seconds to tell their elbow partner what they borrowed. Ask what happens when they borrow something. If you don't hear a student say it, point out that borrowing means you have it temporarily and then give the item back. Conversely, the person giving the item temporarily is lending the item.

Sales Pitch

Explain that one thing that people borrow is money. Ask the following:

- Why might you need to borrow money?
- If you need money to buy something, where might you get that money?

Read *Brody Borrows Money* by Lisa Bullard. Explain that borrowing money sometimes means you have to give back more than you borrowed, and this is called a **fee**. In these activities, the students are going to explore borrowing without fees first, and then borrowing with fees.

The Fine Print

Part 1: Lend and Borrow 10

Explain to the students that they are going to be *borrowing* money (receiving $10) or lending money (giving $10). Match students with a partner, and give each pair (1) a set of Ten Frame cards (See **Resource 8.1: Ten Frame Cards** and preparation instructions); (2) a die; and (3) two copies of **Activity Sheet 8.1: Lend and Borrow.** Assign roles to the partners. One child is the "Lender" and one child is the "Borrower."

Model a first example with the class. Show that students will roll the die the first time to determine how many $10 (full ten frames) they will have. If they roll a 4, each student takes 4 full ten frames. Show students how to count out the forty (like counting $10 bills). The second roll indicates how many $1 amounts each student gets. If they roll a 2, they each take a ten frame card that has two dots. Now each child has 42 represented by ten frame cards in front of them. Ask the following questions (correct answers are in brackets):

- How many tens do you have? [4] How many ones do you have? [2]
- How much money do you have altogether? [$42]

Ask students to record their total in the table in the first row. Then, ask the borrowers to raise their hands. They ask their partner for $10. Ask the Lenders to raise their hands. The Lenders say, "Yes, here is $10." Together they figure out how much money the borrower now has ($10 more) and how much money the lender has ($10 less) and record this in the table.

Repeat the process, and record the new quantity in round 2. You may want to switch roles after round 4. After the students have finished, ask the following:

- What strategies did you use to add 10?
- What strategies did you use to subtract 10?
- What patterns do you notice in your answers? [looking at their tables]

Finally, ask students how they might use the patterns they notice to solve the challenges at the bottom of the page.

Part 2: Figuring Fees

Standards for Mathematical Practice: Students develop MP.8 when they look for patterns for adding or subtracting 10.

A note: This lesson is focused on numbers within 100 (grade 1), but the activity can also be done within 1000 (grade 2), by having the dice signify how many hundreds they will be borrowing (and paying back).

Explain to the students that now they will be borrowing from the bank; in other words, the bank will be lending money—supplying **loans**. Explain that when you borrow money from places like banks, you pay extra. This is called a fee. For example, if you borrow $10, the fee might be $1. Ask the following:

- If you borrow $10, how much will you give back, if the fee is $1?
- If you borrow $20, how much will you need to pay back?

Listen to several students' ideas about how much they might pay back on $20 and the rationales with their answers. Explain that in the lesson today, the bank is charging $1 for every $10 that is borrowed, so they would have to pay back $22 (the $20 they borrowed and one dollar for each ten).

The "Lender" now becomes the "Banker." The Banker charges $1 for *every* $10 that is borrowed. The Borrower begins with rolling the dice. If the roll is a 5, then the Borrower says, "I would like to borrow $50." In their journals, the students should record the borrowed amount ($50). Then, the partners figure out how much has to be paid back to the bank—in this case ($5—$1 for each $10).

After they have had a chance to work with their partners on several more examples, ask the following questions:

- What do you notice about the money you have to pay back?
- What will it cost to borrow $100?

Have students share what strategies they used to figure out how much they had to pay for their fee. Also, ask students what they learned about borrowing money from a business like a bank.

Part 3: The Cost of Borrowing

This third activity is a game, and it combines the ideas of the first two parts. The game is called Pay Back the Bank, and students move up and down the hundreds chart to track their gains and losses of money as they borrow and pay the money back.

Distribute a copy of **Resource 8.2: Bottom-Up Hundreds Chart** to all students, and have them place a counter on 35, representing the money that they have saved. Have them also draw out $35 using the ten frame cards (optional, to provide counting support). (To adapt to within 1000, adapt the chart to count up by 10s to 100 on the bottom row, and so on through 1000 on the top row.) Ask the following questions:

- How much money do you have? How many tens is that? How many ones is that?
- What strategies might you use to add 20 on your hundreds chart? To subtract 20?

If students are not able to add multiples of 10s on the hundreds chart, replace the chart with something more concrete like Unifix cubes, base-10 blocks, or other place-value manipulatives.

Standards for Mathematical Practice: When students notice that adding 10 ones is like adding 1 ten, or a vertical move on the hundreds chart, they are seeing structure in the number system (MP.7).

Explain to students the Pay Back the Bank Game. Directions are provided in figure 8.1.

Pay Back the Bank Game

Materials: Bottom-Up Hundreds Chart (one per player); 1 die

Instructions:

1. Play with a partner or in a group of three.
2. Each player places a counter on 35 of his or her own Bottom-Up Hundreds Chart.
3. Player 1 Borrows: Player 1 rolls the dice and borrows that amount of ten (a roll of 6 means borrow $60). Player 1 would then move his or her counter up 60, up to 95.
4. Player 1 Pays Back: Player 1 figures out what he or she has to pay back. In this case, it is $66 ($60 borrowed + $1 per $10). Player 1 counts back 66 and should now be on 29.
5. Player 2 repeats steps 3 and 4.
6. Repeat, taking turns. Play 10 turns or until one of the players is out of money.

The player with the most money left wins!

Possible Variations: Change the starting number from 35, or change the bank fee to something other than $1 per $10.

Fig. 8.1. Pay Back the Bank game

As the students are playing the game, ask questions that focus on the lesson goals (mathematics and financial literacy). Examples of questions include:

- What patterns are you noticing as you add multiples of 10?
- What patterns are you noticing as you subtract multiples of 10?

- What is the relationship between how much you borrow and how much you pay back?
- What strategies are you using to figure out how much you have to pay back?
- How does the money you have left (on the hundreds board) compare to what you had before you borrowed the money (i.e., what was the start and end for this turn?)?

After students have put their games away. ask them what they noticed as they played. You are listening for ideas related to borrowing and paying back, such as "The more you borrow, the higher the fees," "If you borrow more money, you are going to lose the game," and, "You will lose the game if you roll big numbers because you owe more money to the bank."

Closing the Deal

Share with students that in this lesson they are working on using mental strategies for adding tens. Ask the following questions:

- What did you notice about adding 10 to a number like 24? What about adding a number like 30?
- What did adding 10 look like on the hundreds chart?
- What did you notice about subtracting a 10 from a number? What did it look like on the hundreds chart?

Post an amount of money on the board, such as $44. Ask the following questions:

- How much will I have if I borrow $20?
- If I borrowed it from a friend and then returned the money I borrowed, how much will I have?
- If I borrowed from a business and returned the money, how much will I have?
- What are you learning about borrowing money?
- When might you need to borrow money?

Close by having students write in their journal or say to a partner a pattern they noticed in the lesson that helped them solve the problems or play the game.

Getting Your Money's Worth

This activity can be modified for grade 2 students by focusing on adding hundreds, rather than tens. The loan amount can still be $1 for every $10.

References and Resources

Bullard, Lisa. *Brody Borrows Money.* Minneapolis, Minn.: Millbrook Press, 2013.

Jump$tart Coalition for Personal Financial Literacy. *National Standards in K–12 Personal Finance Education.* Fourth Edition. Second Printing. Washington, D.C.: Jump$tart Coalition for Personal Financial Literacy, 2017.

National Governors Association Center for Best Practices and Council of Chief State School Officers (NGA Center and CCSSO). *Common Core State Standards for Mathematics.* Washington, D.C.: NGA Center and CCSSO, 2010.

Robbins, Tony. *Money Master the Game: 7 Simple Steps to Financial Freedom.* New York: Simon & Schuster, 2014.

Activity Sheet 8.1

Lend and Borrow

ROUND	My Money	Money If I Lend $10	Money If I Borrow $10
1			
2			
3			
4			
5			
6			
7			
8			

Give and Take Challenges:

1. Annie started with $61 and ended with $71. Did she borrow money or lend money?

2. Jacob had some money and lent $10. Now he has $32. What did he have at the start?

3. Natalie had some money and borrowed $10. Now she has $45. What did she have at the start?

Resource 8.1

Ten Frame Cards

Resource 8.1

Ten Frame Cards—page 2

Resource 8.2

Bottom-Up Hundreds Chart

91	92	93	94	95	96	97	98	99	100
81	82	83	84	85	86	87	88	89	90
71	72	73	74	75	76	77	78	79	80
61	62	63	64	65	66	67	68	69	70
51	52	53	54	55	56	57	58	59	60
41	42	43	44	45	46	47	48	49	50
31	32	33	34	35	36	37	38	39	40
21	22	23	24	25	26	27	28	29	30
11	12	13	14	15	16	17	18	19	20
1	2	3	4	5	6	7	8	9	10

Resource 8.3

Number Line *Cut out each line, and tape them end to end.*

Chapter

Learning about Earning

The Bottom Line

Entrepreneurs assume risks and then keep track of earnings and expenses to keep track of the business.

Students form a connection between the real-world aspect of earning money and place value in mathematics. They use a balance sheet to perform mathematical operations based on income and expenses.

Standards Met in This Chapter

CCSSM Standards and Practices

3.NBT.A.2: Fluently add and subtract within 1000 using strategies and algorithms based on place value, properties of operations, and/or the relationship between addition and subtraction.

4.NBT.A.1: Recognize that in a multi-digit whole number, a digit in one place represents ten times what it represents in the place to its right.

4.NBT.B.4: Fluently add and subtract multi-digit whole numbers using the standard algorithm.

5.NBT.A.3: Read, write, and compare decimals to thousandths.

5.NBT.B.7: Add, subtract, multiply, and divide decimals to hundredths, using concrete models or drawings and strategies based on place value, properties of operations, and/or the relationship between addition and subtraction; relate the strategy to a written method and explain the reasoning used.

MP.4: Model with mathematics.

MP.5: Use appropriate tools strategically.

MP.7: Look for and make use of structure.

(National Governors Association Center for Best Practices and Council of Chief State School Officers [NGA Center and CCSSO] 2010)

Jump$tart Standards

Employment and Income: Use a career plan to develop personal income potential.

Standard 2. Compare sources of personal income and compensation.

Knowledge Statements:

Overall:

- Entrepreneurs, who work for themselves by starting new businesses, hope to earn a profit, but accept the risk of a loss if unsuccessful.

Standard 2:

- Identify jobs children can do to earn money.

(Jump$tart Coalition for Personal Financial Literacy 2017)

Future Value

As students get older, some may obtain a part-time job while they go to school. It is important for students to learn to manage their money appropriately, and this chapter lays the foundation for creating a money management system by utilizing a balance sheet to record income and expenses. This chapter can also serve as a precursor for connecting negative numbers to the real world.

The following sections describe the mathematical and financial literacy concepts that this lesson will encompass, as well as the knowledge and experiences that elementary students may bring to the lesson and take from it on its completion.

Mathematics

Why is understanding place value so important in mathematics? Place value is the basis for our number system, and the position of a digit determines its value. Mathematically literate students will be able to make connections between numbers of different place value to compare the size of numbers, perform operations, and bridge the gap when learning about numbers that are only parts of a whole (3.NBT.A.2; 4.NBT.B.4). A big idea for students is to understand that in a multi-digit whole number, each digit has a *place* and a *value*. Through exploration, they will find that a digit in a place represents ten times what it represents in the place to its right (4.NBT.A.1). For example, 500 is ten times more than 50, which is ten times more than 5. It is important for students to recognize this pattern as being the same one carried throughout the place-value system (MP.7).

In order to properly read the standard form of money, students must be able to identify digits to the hundredths place (5.NBT.A.3). When completing operations with money, they must know the procedure for adding, subtracting, multiplying, and dividing with decimals and be able to model this through mathematics with money (5.NBT.B.7; MP.4). In doing so, students will use a balance sheet in order to organize their calculations and data (MP.5).

Did you know? The word "entrepreneur" has evolved in meaning over the years. It was originally an English word loaned from the 18th century French word *entreprende*, which means "to undertake" and was used to describe a manager/promoter of a theatrical production (In the Zone 2018).

Financial Literacy

The monetary system in the United States can be described using place value. One United States dollar equals 10 dimes, which equals 100 pennies. Seeing the connection between place value and money not only helps to build number sense, it helps students to be able to make equal exchanges of money (e.g., ten $10 bills equals $100, and one hundred $10 bills equals $1000).

This chapter also focuses on earning and keeping a record of one's money. Students will be acting as entrepreneurs and will incur income and expenses. Though students may not have a regular paycheck, it is important for them to see the necessity of keeping track of their money and making sure they know when money is entering or leaving their account. By keeping an accurate log of their earnings, students are showing responsibility by being aware of how much money they are saving for a future use.

Student Knowledge and Experience

Students at this age level will have had exposure to place value and place-value patterns. What students might not have mastered is place value with decimals and performing operations with multidigit decimals to hundredths. If students have not yet learned about decimals, that is okay for this lesson. Simply explain to them that money is always written out to the hundredths place. Show an example. Many students at this level have seen money written out in decimal notation, so talking about decimals and money actually makes an excellent real-world connection.

Lesson Plan

Learning Targets

- Recognize that the digits of a number have a place and value.
- Identify that a digit has a value of ten times more than the same digit to its right.
- Perform mathematical operations using whole numbers and decimals.
- Use a system of record keeping to log incoming and outgoing money.

Resources and Tools

- **Activity Sheet 9.1:** Balance Sheet (one for each student)
- **Resource 9.1:** Money and Place Value (one for each student)
- **Resource 9.2:** Task Cards 1–3 (class set, or a set for each pair of students)
- *Once upon a Company . . . A True Story* by Wendy Anderson Halperin (1998) (or another book about entrepreneurship)
- Base-10 blocks (optional)

Preparation Prior to Lesson

Sales Pitch: Obtain a copy of *Once upon a Company . . . A True Story* by Wendy Anderson Halperin.

Part 1: Prepare enough copies of Resource 9.1 for each student

Part 2: Prepare enough copies of Activity Sheet 9.1 as needed for each student. Prepare a class set or partner sets of Resource 9.2 as needed.

Children's Literature Option

See the Literature for Financial Literacy list on p. 203 for additional relevant books to incorporate into the lesson.

Key Language

earn: Receiving money for doing work.

income: Money that is received; for example, earned or as a gift.

entrepreneur: A person who starts or organizes a business and understands that there are risks concerning earnings and losses.

Money Talks

Begin by talking with students about the different ways they receive money. This is called **income** (money coming in). Frequently their money will come from gifts they are given, but sometimes children do various jobs to **earn** their own money. Ask students the following:

- Have you ever earned money for doing work?
- What did you do with the money once you earned it?

Explain to students that they are going to learn about earning money, how to record money amounts, and how to keep track of income and outflow of a business.

Sales Pitch

Share with students the book *Once Upon a Company . . . True Story* by Wendy Anderson Halperin. As you read the book, consider asking questions, such as the following:

- Why did the children want to earn money?
- What is an **entrepreneur**?
- Were the children successful?

Explain to students that they will be exploring place value through money and will have the opportunity to start their own business using the following activities.

The Fine Print

Part 1: Money and Place Value

Standards for Mathematical Practice: As students learn to represent monetary amounts using decimal notation, they are then able to write expressions and equations to represent situations involving money (MP.4).

Remind the students that they've been working with numbers that have different values based on their position, or place. As an extension of this, we will explore money and how it relates to place value. Ask students the following:

- How much is a penny worth?
- How do I write that in decimal notation? [$0.01]

It is important when writing with money to use proper and consistent format. Show students the written difference between $0.01 and 1¢. Frequently, students have a misconception about written notation and will write $0.01¢. Explain to students that the cent sign is not needed when the dollar sign is used.

Next, say to the students, "Now that you know how to write 1 cent as $0.01, how many pennies make a dime, and how can we model the mathematics?" Some students might show $0.01 × 10, while others might choose to add $0.01, ten times. As they answer, write this comparison on the board (10 pennies = 1 dime).

Standards for Mathematical Practice: Students develop MP.1 as they use the mathematics to make sense of the problem and figure out how long it will take to save the money for their purchase.

Then have students predict how to write ten cents in decimal notation. Ask why they think this and model answers on the board. Verify with students how to write ten cents in decimal notation [$0.10]. Next, ask how many dimes there are in a dollar. Write this comparison [10 dimes = 1 dollar]. Finally, ask students how they would write $1.00 in decimal notation. Ask the following:

- Why is the 1 to the left of the decimal point?
- What do the numbers to the right of the decimal point mean?
- What do you notice about each digit value and the digit value to the right? [The digit value is 10 times more than the digit value to its right.]
- Does this pattern hold true for any number?

After practice with writing money in decimal notation, students should be able to make a connection that one penny is one hundredth of a dollar, and it can be represented by $0.01, where the 1 is in the hundredths place. A dime is worth ten cents, and it can be represented by one tenth of a dollar, or $0.10.

Standards for Mathematical Practice: Students are developing MP.7 as they start to recognize the pattern in place value and are able to apply that knowledge to calculations with money.

Using **Resource 9.1: Money and Place Value** as a recording page, show different amounts of money (holding it up or projecting it on a screen) and invite students to record the quantities. After each example, invite students to compare how they recorded their quantity with others.

To follow up with each example, ask questions such as the following:

- If you earned twenty dollars more, how would your number change? Stay the same?
- If you found 22 dimes, and added it to your total, how much would you have?
- If you spent 30¢, how would your quantity change?
- How many dimes does it take to increase your quantity by $10? How many $10 amounts does it take to increase your quantity by $100?

Part 2: Earning and Entrepreneurs

Referring back to the Money Talks section, ask students for some ways that they get money. Some answers might be birthdays, religious milestones, doing chores around the house, or walking a neighbor's pet. Ask students what do they do with the money when they receive it. Explain that this is *income.* Remind them from the book you have read how The College Fund Wreath Company was started and how it was managed.

In this activity, the students will explore being entrepreneurs. Remind students that entrepreneurs organize a business and understand that there are risks concerning earnings and losses. As entrepreneurs, they are going to hypothetically open a dog-walking business. Ask, "What risks might you encounter in this business? What losses and expenses might you have? What earnings might you receive?"

Place students in small groups. Their first order of business is to hash out the details of their dog-walking business. They need to decide on a name for their business and on how much they will charge per dog and per minute. (If time allows, add a visual arts component by having your students make their own advertising flyer based on their business details.) Once decisions have been made, explain to the students that they will need to keep track of the money earned or spent based on the card from **Resource 9.2: Task Cards** card they drew from a stack (or other cards of your creation). To keep a record, they are going to maintain a balance sheet. (Pass out copies of **Activity Sheet 9.1: Balance Sheet**). Any money that goes in or out of the account must be recorded on this sheet. See figure 9.1 as an example.

As students work, remind them of the importance of lining up the place values and decimal points when adding and subtracting. If you have not taught multiplication with decimals, explain to the students that you are multiplying as though you are being paid in pennies (with whole numbers). When you calculate your final answer, it is like turning pennies into dollars (which is what they practiced in part 1.) Lined or gridded paper can also assist in keeping numbers aligned.

Balance Sheet *We decided on 25¢ per dog per minute.

Shuffle and place task cards face down in front of you. Draw 5 task cards from the pile, one at a time. As you choose a card, record the task description and the amount of money you earned or spent. You may need to do some pre-calculations before you add it to your balance sheet. These can be done below. Keep a running total in the last column. Your final total should be the same as the amount of money you calculated in the Total Money column of line 5.

Card Description	Money Earned	Money Lost	Total Money
1. Finn – 1 dog/40 min.	$10.00		$10.00
2. Milo + Daisy – 2 dogs/25 min	$12.50		$22.50
3. Bought Ball		$1.25	$21.25
4. Otto – 1 dog/15 min	$3.75		$25.00
5. Ice Cream with Guns		$3.00	$22.00
Final Total:			$22.00

Show your work! For each card, record your calculations below.

1.	2.	3.
40 × 25; 200 + 800 = 1000 = $10.00	25 × 25; 125 + 500 = 625; 625 × 2 = 1250 = $12.50	22.50 − 1.25 = 21.25
4. 15 × 25; 75 + 300 = 375 = $3.75; 21.25 + 3.75 = $25.00	**5.** 25.00 − 3.00 = $22.00	**Final** $22.00

Fig. 9.1. Example of a balance sheet

Closing the Deal

Once your students have had time to create one or more balance sheets, come back together and revisit the questions posed before they explored the business (about the risks a dog-walking business has). In addition, ask the following questions:

- Did you earn money from your dog-walking business? Why or why not? How might you change your business if you were really going to have a dog-walking business?
- What are some things you have to know in order to keep an accurate balance sheet? (Answers might include the following: line up decimals, money that is lost is subtracted, money that is earned is added, sometimes we have to pay money, it can be hard to keep track of money.)
- If you walked 5 dogs for 2 hours at $7.00 an hour, how much would you make?

Invite groups to brainstorm other entrepreneurial ideas that they could do at their age. Ask:

- What are some things you could do with that earned money?

Responses might include tutoring, yard work, cleaning, or other creative ideas. After they settle on an idea, have the group prepare a business brief, telling (1) the business idea, (2) possible risks, (3) anticipated/possible earnings, and (4) expenses.

In sharing their business ideas, challenge students with questions that incorporate place value concepts, such as these examples:

- If it takes one-half of a day to earn $10, how long will it take to earn $100? $1000?
- If your idea involves a cost of one dime, how many dimes will you need to earn $1? $10? $100?
- Which is better, offering a product for $15 that 10 people will purchase, or offering a product for $5 that 100 people will purchase?

Getting Your Money's Worth

This activity can be modified for a span of grade levels and abilities by manipulating the task cards, earnings, or balance sheet. Decimals may be left out of the activity by only including whole numbers in the wages and calculations. This is also a good activity for teaching students responsibility for their money (and dogs), as well as emphasizing the importance of saving for the future. To challenge students, give them further financial problems related to a balance sheet.

References and Resources

Halperin, Wendy Anderson. *Once upon a Company . . . A True Story.* New York: Orchard Books, 1998.

Jump$tart Coalition for Personal Financial Literacy. *National Standards in K–12 Personal Finance Education.* Fourth Edition. Second Printing. Washington, D.C.: Jump$tart Coalition for Personal Financial Literacy, 2017.

National Governors Association Center for Best Practices and Council of Chief State School Officers (NGA Center and CCSSO). *Common Core State Standards for Mathematics.* Washington, D.C.: NGA Center and CCSSO, 2010.

Activity Sheet 9.1

Balance Sheet

Shuffle and place task cards face down in front of you. Draw 5 task cards from the pile, one at a time. As you choose a card, record the task description and the amount of money you earned or spent. You may need to do some calculations before you add it to your balance sheet. They can be done below. Keep a running total in the last column. Your final total should be the same as the amount of money you calculated in the Total Money column of line 5.

Card Description	Money Earned	Money Lost	Total Money
1.			
2.			
3.			
4.			
5.			
Final Total:			

Show your work! For each card, record your calculations below.

1.	2.	3.
4.	5.	Final

Resource 9.1

Money and Place Value

Record the amount of money represented in the pictures you see.

$						
$						
$						
$						
$						
$						
$						
$						
	Thousands	Hundreds	Tens	Ones	Tenths	Hundredths

Resource 9.2

Task Cards 1

Walked Rufus for 30 minutes. 	Walked Sadie for 25 minutes. 	Walked Maggie for 45 minutes. 
Broke Sophie's leash. Pay $7.50.  	Bought Boomer a new ball. Pay $1.25. 	Bought one bone and one treat for Lincoln. Pay $0.75 for each. 
Walked Otto for 15 minutes. 	Walked Bailey twice today, once for 20 minutes and the other time for 25 minutes. 	Walked Finn to the park and back for a total of 40 minutes. 

Resource 9.2

Task Cards 2

Walked Carly for 50 minutes. 	Walked Milo and Daisy for 50 minutes. 	Walked Sammy twice today, once for 40 minutes and once for 20 minutes. 
Bought Buster a chew toy. Pay $3.00. 	Walked Zoe around the block for 20 minutes. 	Bought one treat each for Jake and Sasha. Pay $0.60 for each. 
Walked Bella for 30 minutes. 	Walked Jasmine to the dog park and back for a total of one hour. 	Lost Oliver's kerchief. Pay $2.00. 

Resource 9.2

Task Cards 3

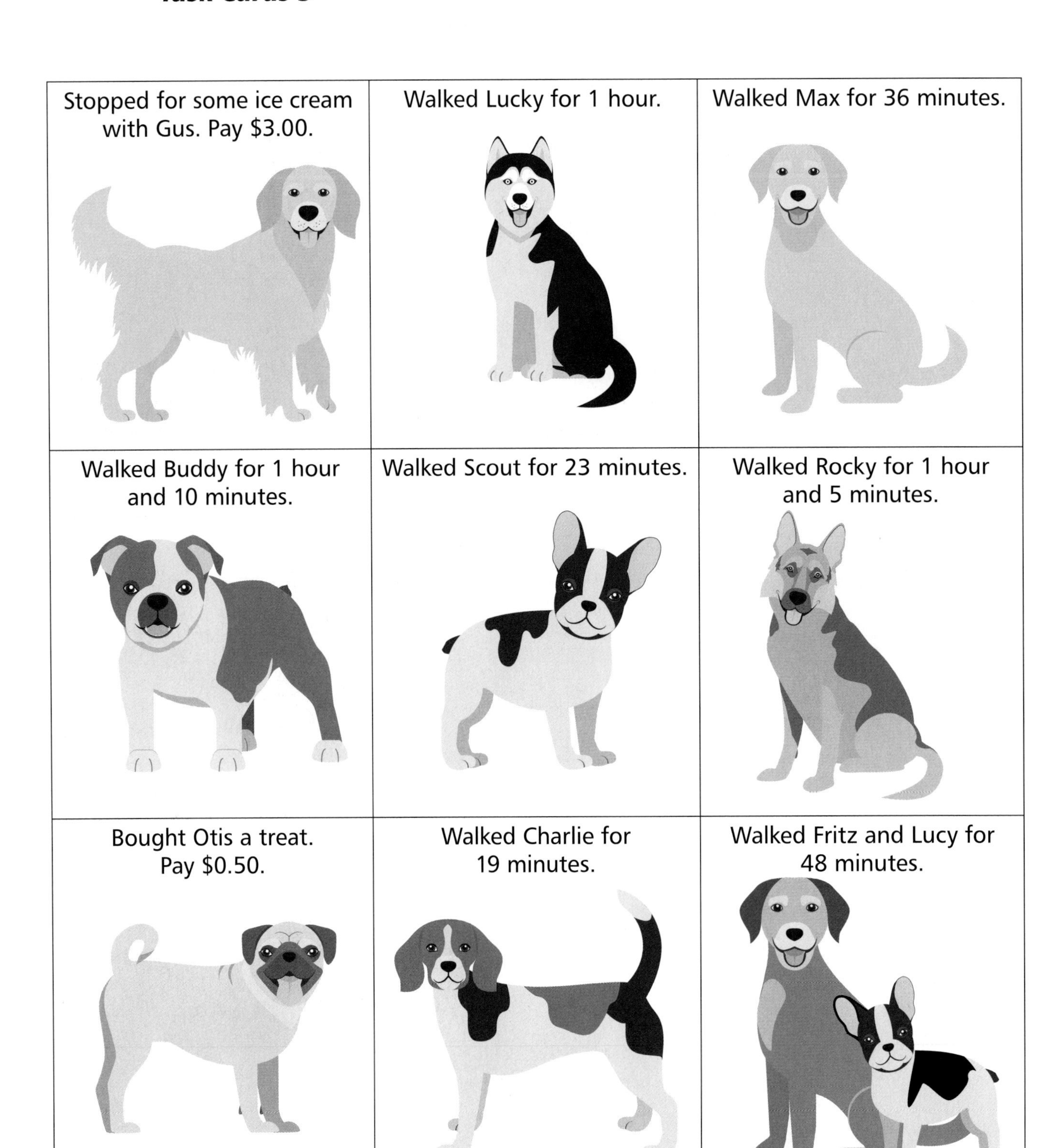

Stopped for some ice cream with Gus. Pay $3.00.	Walked Lucky for 1 hour.	Walked Max for 36 minutes.
Walked Buddy for 1 hour and 10 minutes.	Walked Scout for 23 minutes.	Walked Rocky for 1 hour and 5 minutes.
Bought Otis a treat. Pay $0.50.	Walked Charlie for 19 minutes.	Walked Fritz and Lucy for 48 minutes.

Chapter

Paychecks and Payments

The Bottom Line

Gross pay and net pay differ based on various withholdings.

Students explore the difference between gross and net pay by examining and calculating wages using a pay stub. They learn that they might not always take home all of the money they initially earned.

Standards Met in This Chapter

CCSSM Standards and Practices

4.NBT.B.4: Fluently add and subtract multi-digit whole numbers using the standard algorithm.

4.NF.C.6: Use decimal notation for fractions with denominators 10 or 100.

5.NBT.B.7: Add, subtract, multiply, and divide decimals to hundredths, using concrete models or drawings and strategies based on place value, properties of operations, and/or the relationship between addition and subtraction; relate the strategy to a written method and explain the reasoning used.

5.NF.B.6: Solve real world problems involving multiplication of fractions and mixed numbers, e.g., by using visual fraction models or equations to represent the problem.

MP.4: Model with mathematics.

MP.6: Attend to precision.

(National Governors Association Center for Best Practices and Council of Chief State School Officers [NGA Center and CCSSO] 2010)

Jump$tart Standards

Saving and Spending: Apply strategies to monitor income and expenses, plan for spending and save for future goals
Knowledge Statements:
Overall:

- People are required to pay taxes, for which they receive government services.

Employment and Income: Use a career plan to develop personal income potential.

Knowledge Statements:

Overall:

- People can earn income in the form of wages, salaries, and tips.

Standard 2:

- Compare sources of personal income and compensation.
- Explain the difference between a wage and a salary.

(Jump$tart Coalition for Personal Financial Literacy 2017)

Future Value

Understanding the concept of gross versus net pay is critical so that students begin asking questions about employment opportunities. When students become older, the difference between their gross and net pay may become larger and larger. A common misconception can occur when an employee applies for a job and the employer mentions the wage or salary that the employee will receive. Employees sometimes believe that this is their net pay, when typically this amount is their gross pay.

The following sections describe the mathematical and financial literacy concepts that this lesson will encompass, as well as the knowledge and experiences that elementary students may bring to the lesson.

Mathematics

This chapter focuses on adding, subtracting, and multiplying whole numbers and decimals (4.NBT.B.4; 5.NBT.B.7). For students not ready for decimals, these activities can still be accomplished using only whole-number values by rounding to the nearest dollar. The teacher may supply the deduction amount, and the student can still find the difference between the gross pay and the net pay. For those students ready for decimals, students will determine percentages written as fractions out of 100 (or the decimal equivalent) and then apply this information to their calculations (4.NF.C.6; 5.NF.B.6). Throughout the activities, students will be modeling mathematics in everyday life by writing equations to describe situations (MP.4). Students will also need to attend to precision when figuring their calculations, deciding if they will round to the nearest dollar or penny (MP.6). You don't want to be wrong when determining how much you will get paid!

Financial Literacy

This activity provides students the chance to compare net pay and gross pay and to begin a discussion about taxes that are deducted from a paycheck. When students first get a jobs such as occasional babysitting or lawn mowing, they are casual employees and usually do not have to pay taxes. But when they are hired more regularly by a business, they will have taxes and perhaps other fees taken out of their paycheck. For example, their first part-time job might be a paper route or working at a fast-food restaurant, and taxes will be withheld from their paychecks. Taxes pay for many things, from schools to roads. Taxes can be collected in a variety of ways, including personal income (what you earn), property tax (what you own), and sales tax (what you buy). In this lesson, the focus is on personal income tax and how that affects a person's net pay. An understanding of gross versus net pay will help students create a budget, which in turn will help with how they think about spending and saving money.

Student Knowledge and Experience

Because taxes focus on percentages, and students' experiences are based on fractions out of 100, and related decimals, students will need concrete and visual representations of percent. For this lesson, they just need to know that taxes are a rate out of 100. So, a tax rate of 2.0 or 2 percent means that if they earned $100, they would pay $2 to the government and keep $98. A hundreds chart or base-10 blocks can help students to see this relationship.

Students might not have had much exposure to income taxes, but they may know about sales tax. Explain the different ways the government taxes its citizens. Explain that we don't get to keep all the money we earn. It is our responsibility as citizens of the United States to contribute to our city, state, and federal governments by paying taxes, which are used to provide services to the people.

Lesson Plan

Learning Targets

- Use addition, subtraction, and multiplication to calculate gross and net pay.
- Explain the difference between gross pay and net pay.
- Write percentages in fraction form as a quantity out of 100, and use fractions to solve problems.
- Explain why we pay taxes in the United States to local, state, and national governments.

Did you know?
According to an article by Lenhardt (2017), in the United States, "Everyone who earns a paycheck pays a federal income tax. Forty-three of the 50 states charge their citizens an income tax. The seven states that do not have a state income tax are Alaska, Florida, Nevada, South Dakota, Texas, Washington, and Wyoming."

Resources and Tools

- **Activity Sheet 10.1:** Estimated Tax Rates: What Might You Pay?
- **Activity Sheet 10.2a or 10.2b**: Pay Stubs, or Pay Stubs with Overtime (one for each student)
- **Resource 10.1:** Pay Stub Example (one to project for the class)
- *Neale S. Godfrey's Ultimate Kids Money Book* by Neale S. Godfrey (1998) (or another book about gross and net pay)
- Calculators (optional)

Preparation Prior to Lesson:

Money Talks: Project a copy of Resource 10.1 for all students to see.

Sales Pitch: Obtain a copy of *Neale S. Godfrey's Ultimate Kids Money Book* by Neale S. Godfrey.

Part 1: Display the website for the Bureau of Labor Statistics *Occupational Outlook Handbook* (https://www.bls.gov/ooh/home.htm). Prepare enough copies of Activity Sheet 10.1 for each student.

Part 2: Prepare enough copies of Activity Sheet 10.2a, and 10.2b for each student.

Children's Literature Option:

See the Literature for Financial Literacy list on p. 203 for additional ideas.

Key Language

gross pay: Amount of money earned before taxes are taken out.

net pay: Amount of money earned after taxes are take out.

salary: Compensation for work, expressed as an annual sum and paid in prorated portions regularly—usually weekly, biweekly, or monthly.

tax: Amount of money removed from gross pay to be used by the government and other entities.

wage: Compensation for work, usually calculated on an hourly, daily, or piecework basis and paid on schedule—usually weekly, biweekly, or monthly.

Money Talks

Share a pay stub with the class (**Resource 10.1: Pay Stub Example**). Ask students to look at what the person made (**gross pay**) and then the size of the payment (**net pay**). You might present this as a missing-value scenario: Gross Pay – ______ = Net Pay. Ask students the following questions:

- What might be listed in the blank?
- What fees does a person pay? [tax, etc.]
- Which fees are optional? Which are required?
- Why do you think we pay **taxes**?

List their ideas on the board. Note: This pay stub and these questions will be revisited in part 2 of the lesson.

Sales Pitch

Neale S. Godfrey's Ultimate Kids Money Book (1998) is an excellent resource for learning about financial literacy. Read pages 94–97 with the class to gain some insightful information on why we pay taxes, how tax is collected, how tax percentages are determined, sales tax, income tax, and who pays taxes. According to Godfrey (1998), answers would includc thc following:

- Why we pay taxes—The government uses taxes to pay for the services it provides, such as schools, libraries, some hospitals, roads and bridges, and the armed forces.
- How tax is collected—The government collects taxes when you buy certain goods, from income people make, from property owners, or from large inheritances.
- How tax percentages are determined—The government looks at how much money it needs and how much people can afford to decide how much money is collected in taxes.
- What is sales tax?—Sales tax is a tax on sales and some services that is usually added to the price by the seller.
- What is income tax?—A tax on a person's income determined by how much you earn.
- Who pays taxes?—Everyone who lives and works in the United States must pay taxes. There are a few exceptions, and these people or organizations are considered tax-exempt.

A part of becoming financially literate is understanding why we pay taxes and where that money is used.

The Fine Print

Part 1: Net Pay versus Gross Pay

Standards for Mathematical Practice: When students are performing calculations such as taking a percentage of their gross pay to be withheld by the federal government, they will need to use precision to ensure their calculations are correct and that they withhold enough money to pay their taxes! (MP.6).

To begin, explain to students that taxes are set up so that people who earn a lot of money pay a higher amount for every $100, then people who earn less. Begin with a few examples, such as: *Christopher earned $3000.* Explain that his tax rate is 1/10. Ask students to share with a partner how to record this as a decimal [0.1]. Ask students to use either the fraction or the decimal to figure out what one-tenth of $3000 is. Share strategies and answers. Explain that $3,000 is Christopher's gross pay, $300 is the tax he owes, and therefore his net pay (take home) is $2,700. Then, ask students what Christopher would pay if his tax rate was three-tenths. Repeat the process.

Distribute **Activity Sheet 10.1: Estimated Tax Rates: What Might You Pay?** and ask students to work with a partner to figure out each person's income tax and their net pay. After students have completed the activity sheet, ask them to share and compare strategies they used for finding the various tax rates.

Next, ask students if they have ever thought about what they might want to be when they grow up. Point out that while pay is not the only thing to consider in selecting a job, it is one of the factors people consider. If students have access to technology, invite them to spend time exploring different career options and possible annual **salary** using the *Occupational Outlook Handbook* put out by the Bureau of Labor Statistics (https://www.bls.gov/ooh/home.htm). Encourage them to identify three to five careers that sound interesting. It would be a good idea to consider a diverse range of occupations and to demonstrate that there are many, many career options. Once they have created a list, along with the anticipated annual salary, have them return to Activity Sheet 10.1 and use the table of estimated taxes to figure out what taxes they would pay and what their net pay would be. Alternatively, have students record career ideas on sticky notes early in the lesson, and as they are doing Activity Sheet 10.1, look up the estimated salaries for these careers and post or project those values for students.

At this point, students have an annual salary and know the amount of their gross pay that will be withheld for taxes. Ask students to write an equation to represent their net pay. Give students time to explore what operation they will be using and to model the mathematics. Depending on the grade level or interest of students, this would be a good time to extend the lesson to a monthly budget for saving, spending, and sharing.

Part 2: Interpreting Paychecks

Standards for Mathematical Practice: Students develop MP.4 as they write equations that model a general process for calculating taxes and net pay.

Explain that in the first activity, they were exploring different salaries. Salaries tell what you earn in a year, and paychecks are divided by 12 (monthly) or 26 (biweekly). But most people initially work for hourly **wages**. Ask students to name jobs that they think are paid hourly and ones that they think are paid with a salary. Explain that now they are going to explore paychecks based on an *hourly wage*. Refer back to Resource 10.1 as an example of a pay stub. Resource 10.1 may have more detail than your students need; depending on academic level and ability, students might wonder what Social Security is, and the differences between state and federal taxes. There might be some other terminology that students are not familiar with, such as YTD (year to date), and they may wonder why YTD is the same as current amount and withholdings. This pay stub is an example of the first earnings statement of the new year. After this, the YTD amounts would increase throughout the year. Have students examine the pay stub, and give them time to explore what the different categories might mean or how the amount was calculated. Once students have had time to look, ask questions such as the following:

- About what fraction (decimal) of Darnell's gross pay goes to taxes?
- What are some voluntary deductions made by Darnell?
- What fraction (decimal) of Darnell's gross pay is used for voluntary deductions?
- Why might Darnell be saving $15 from each paycheck?

There are two ways to use Activity Sheet 10.2a and 10.2b. You can do both, starting with 10.2a and moving to 10.2b, which includes overtime pay. Or, you can differentiate assigning students to one of the two, or letting them choose. Explain that they will have the opportunity to calculate their own pay stub. Make enough copies of **Activity Sheet 10.2a or 10.2b: Pay Stubs** for every student. Provide student with an hourly rate or have them look up a reasonable rate for your city or town. Begin with having them calculate their salary for 10 hours for one week. Ask students how they would calculate the hourly total, overtime total (if applicable), and gross pay. (Hourly total is number of hours worked multiplied by hourly rate). Repeat for a full-time job (40 hours). For students using 10.2b, explore for a 50-hour week. Ask, "What strategies can you use to calculate an overtime rate of 1.5? (Overtime total is number of overtime hours worked multiplied by overtime rate.) To calculate time and a half, multiply your hourly rate by 1.5 or figure half of your hourly rate and add it to your regular hourly rate which equals time and a half.)

Standards for Mathematical Practice: Calculating taxes and determining net pay is an excellent way to represent a real-world mathematics problem by applying the correct mathematical model (MP.4).

Explain to students that figuring our net pay (take-home pay) is often done through estimation. Have students think of a way they can mentally figure out how much they will get to "take home" if they know how much they have earned. Have the students model the mathematics by creating equations to describe the pattern they notice between gross pay and take-home pay. (These can vary, but should result in a reasonable estimate.)

Closing the Deal

Through this lesson, students can begin to understand that when they earn income for work, sometimes they don't keep all of their money. Use a class chart with students to brainstorm (1) instances when students might need to pay taxes and (2) how they benefit from taxes that get paid. Some examples of when they (or their families) pay taxes might be buying clothes, groceries, house, gas, and so on. Ways they benefit from taxes include their schools and (yes) their teachers.

Finally, ask students to prepare a picture or an explanation that would help someone who just got a job earning $20,000 to figure out how much money he or she will actually be bringing home. Invite students to share their pictures and explanations.

Getting Your Money's Worth

This lesson is a great first step in helping students understand that when they earn income, that income is often taxed, and students will receive less money than expected. This can cause problems if they don't budget accordingly for their expenses and consequently cannot pay their monthly bills. Additionally, students can consider what will happen with earnings over time, by creating a monthly budget or calculating what their net pay will be over the summer or other period of time. This provides an excellent context for computing with whole numbers and decimals as well as for helping students see the time that is involved in saving one's earnings. Finally, this chapter applies knowledge of fractions and decimals to explore the differences between net and gross pay, including what is involved in calculating taxes. But these ideas extend well beyond taxes. In fact, rates are one of the biggest ideas in mathematics. You can ask your students to think about other situations where they must figure out a part of a whole, such as sale prices.

References and Resources

Bureau of Labor Statistics (U.S. Department of Labor). *Occupational Outlook Handbook* [website]. https://www.bls.gov/ooh/home.htm.

Godfrey, Neale S. *Neale S. Godfrey's Ultimate Kids' Money Book*. New York: Simon & Schuster, 1998.

Jump$tart Coalition for Personal Financial Literacy. *National Standards in K–12 Personal Finance Education.* Fourth Edition. Second Printing. Washington, D.C.: Jump$tart Coalition for Personal Financial Literacy, 2017.

Lehnardt, Karin. "70 Interesting Facts about Taxes" (2017). FactRetriever.com [website]. https://www.factretriever.com/taxes-facts.

National Governors Association Center for Best Practices and Council of Chief State School Officers (NGA Center and CCSSO). *Common Core State Standards for Mathematics*. Washington, D.C.: NGA Center and CCSSO, 2010.

Activity Sheet 10.1

Estimated Tax Rates: What Might You Pay?

The table provides general rates for what you might pay, based on the number of people in your household and how much you earn. Compute the decimal equivalence for each tax rate.

Salary	Tax Rate (fraction form)	Tax Rate (decimal form)
$0 to $9,400	1/10	
$9,401 to $40,000	15/100	
$40,001 to $190,000	3/10	
$190,001 to $420,000	35/100	
$420,001+	4/10	

Use the fractions or decimals above to calculate each person's estimated federal income tax, and their net pay.

Worker	Gross Pay	Estimated Taxes	Estimated Net Pay
Amalia	$8,600		
Caroline	$50,000		
Johanna	$500,000		
Miguel	$90,000		
Jeremy	$35,000		
Rebecca	$250,000		
Marie	$65,000		

Activity Sheet 10.2a

Pay Stubs

<table>
<tr><th colspan="5">On the Money, Inc.</th></tr>
<tr><td colspan="2">Employee Name:</td><td colspan="2">Employee ID:</td><td>Date:</td></tr>
<tr><td colspan="2">Number of Hours Worked:</td><td colspan="2">Hourly Rate:</td><td>Hourly Total:</td></tr>
<tr><td>Gross Pay:</td><td colspan="2">Tax Rate:</td><td>Optional Deductions:</td><td>Net Pay:</td></tr>
</table>

Show your work here:

<table>
<tr><th colspan="5">On the Money, Inc.</th></tr>
<tr><td colspan="2">Employee Name:</td><td colspan="2">Employee ID:</td><td>Date:</td></tr>
<tr><td colspan="2">Number of Hours Worked:</td><td colspan="2">Hourly Rate:</td><td>Hourly Total:</td></tr>
<tr><td>Gross Pay:</td><td colspan="2">Tax Rate:</td><td>Optional Deductions:</td><td>Net Pay:</td></tr>
</table>

Show your work here:

Activity Sheet 10.2b

Pay Stubs with Overtime

<table>
<tr><th colspan="4">On the Money, Inc.</th></tr>
<tr><td>Employee Name:</td><td colspan="2">Employee ID:</td><td>Date:</td></tr>
<tr><td>Number of Hours Worked:</td><td colspan="2">Hourly Rate:</td><td>Hourly Total:</td></tr>
<tr><td>Number of Overtime Hours:</td><td colspan="2">Overtime Rate: 1.5</td><td>Overtime Total:</td></tr>
<tr><td>Gross Pay:</td><td>Tax Rate:</td><td>Optional Deductions:</td><td>Net Pay:</td></tr>
</table>

Show your work here:

<table>
<tr><th colspan="4">On the Money, Inc.</th></tr>
<tr><td>Employee Name:</td><td colspan="2">Employee ID:</td><td>Date:</td></tr>
<tr><td>Number of Hours Worked:</td><td colspan="2">Hourly Rate:</td><td>Hourly Total:</td></tr>
<tr><td>Number of Overtime Hours:</td><td colspan="2">Overtime Rate: 1.5</td><td>Overtime Total:</td></tr>
<tr><td>Gross Pay:</td><td>Tax Rate:</td><td>Optional Deductions:</td><td>Net Pay:</td></tr>
</table>

Show your work here:

Resource 10.1

Pay Stub Example

Darnell's Grass Cutting — Earnings Statement

Employee Name	Social Security No.	Begin date	End date	Check date
Darnell Woodward	xxx-xx-9861	10/15/17	10/22/17	10/25/17

Earnings	*Hours*	*Rate*	*Current amount*	*Withholdings*	*Current amt*	*Year to date ($)*
Gross Pay	6.00	$8.00/hr	$48.00	Federal Tax	3.68	3.68
				State Tax	2.35	2.35
				Social Security	0.81	0.81
				Savings deposit	15	15
				Chariable donation	5	5

Current amount	Current withholdings	Net pay	YTD earnings	YTD withholdings	YTD net pay	Check no.
$48.00	$26.84	$21.16	$48.00	$21.16	$21.16	1234

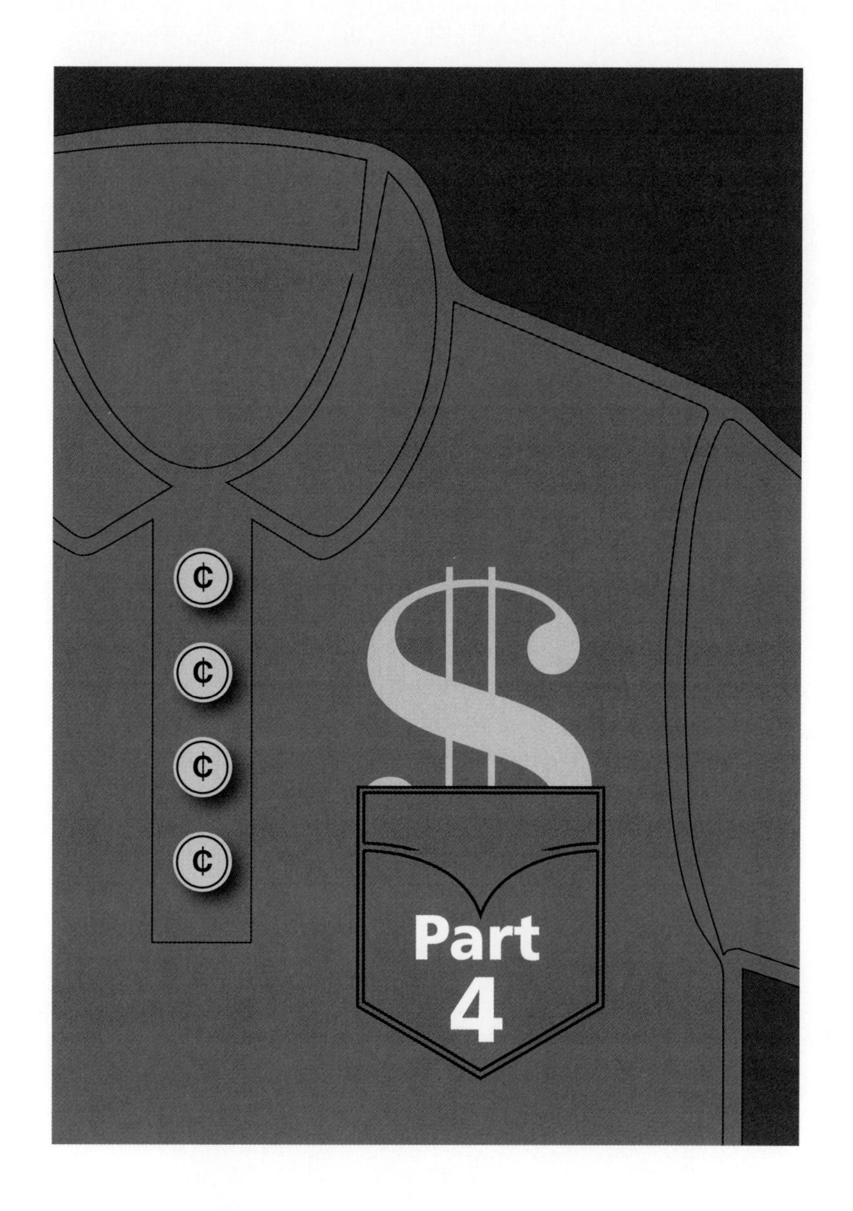

Number and Operations—Fractions

Chapter

What's My Rule for Saving Money?

The Bottom Line

Investing money may help in reaching future financial goals.

Students investigate patterns through in-and-out tables as well as mentally solve scenarios involving money. They will also be introduced to the simple-interest formula, and they will model how to convert a percent into a decimal.

Standards Met in This Chapter

CCSSM Standards and Practices

4.OA.C.5: Generate a number or shape pattern that follows a given rule. Identify apparent features of the pattern that were not explicit in the rule itself. *For example, given the rule "Add 3" and the starting number 1, generate terms in the resulting sequence and observe that the terms appear to alternate between odd and even numbers. Explain informally why the numbers will continue to alternate in this way.*

4.NF.C.6: Use decimal notation for fractions with denominators 10 or 100.

5.OA.B.3: Generate two numerical patterns using two given rules. Identify apparent relationships between corresponding terms. Form ordered pairs consisting of corresponding terms from the two patterns, and graph the ordered pairs on a coordinate plane. *For example, given the rule "Add 3" and the starting number 0, and given the rule "Add 6" and the starting number 0, generate terms in the resulting sequences, and observe that the terms in one sequence are twice the corresponding terms in the other sequence. Explain informally why this is so.*

MP.4: Model with mathematics.

MP.7: Look for and make use of structure.

MP.8: Look for and express regularity in repeated reasoning.

(National Governors Association Center for Best Practices and Council of Chief State School Officers [NGA Center and CCSSO] 2010)

Jump$tart Standards

Investing: Implement a diversified investment strategy that is compatible with personal financial goals.

Standard 1. Explain how investing may build wealth and help meet financial goals.

Knowledge Statements:

Overall:

- Investing means putting money to work to earn more money for the future.
- People invest for future and financial goals.

Standard 1: Calculate simple interest earnings.

(Jump$tart Coalition for Personal Financial Literacy 2017)

Future Value

Investing money is one of the top concepts in financial literacy, a topic from which both students and adults can benefit. Our youngest learners aren't ready to begin investing in the stock market, but they are ready to learn how to make their money grow. This chapter is an introduction to functions through the use of in-and-out tables and the simple-interest formula.

The following sections describe the mathematical and financial literacy concepts that this lesson will encompass, as well as the knowledge and experiences that elementary students may bring to the lesson and take from it on its completion.

Mathematics

This chapter is divided into parts. Each part can be used independently of the others over the span of a unit. Students begin by exploring percents and how each percent is one of one hundred. They will model this knowledge by creating visual representations using a hundred square (MP.4). Students will then apply this understanding to explore the relationship between percents and decimals (4.NF.C.6). Even though a percent is written as a "whole number," it is still just a part of 100. Students will look for and make use of structure when finding 10 percent of a number or 1 percent of a number (MP.7).

Next, students will be asked to identify patterns using in-and-out tables. It is important for students to practice the skill of recognizing relationships between numbers and finding the rule that applies to the situation (4.OA.C.5; 5.OA.B.3; MP.8).

Financial Literacy

Students may have experiences with simple investing from an early age. Many students will have a simple interest savings account over the course of their lifetime, and by starting early, students can watch their money grow as they age. The thought of money "growing" can be an abstract concept for many of our youngest learners, because they tend to not think about the future that is more than a day or two away. Investing becomes increasingly more important, though, as students grow older and begin thinking about buying large-ticket items such as a vehicle and about paying for college. The earlier students consider investing, the more power their money will have as it grows over time.

Did you know?
More than *one half* (55 percent) of Americans are at risk for being unprepared to retire, according to Fidelity's retirement readiness survey (2016).

Student Knowledge and Experience

This lesson builds on and supports students' emerging understanding of place-value concepts and mental computation. Some students will have difficulty with calculating 1 percent or 10 percent of a number and with connecting how and why the decimal place shifts. Providing a visual will help students connect this concept to real-world experiences they may have encountered.

Students most likely will not have any experience with investing. Though some may have started a savings account with their guardians, many may not have seen a statement or had the opportunity to watch their money "grow." Because investing money may be a foreign concept to students, providing concrete examples through in-and-out tables will help them visualize the simple interest model.

Lesson Plan

Learning Targets

- Introduce percentages.
- Use percentages to calculate simple interest, using a numerical representation.
- Determine the rule of a set of numerical values in a table.
- Acknowledge interest as a way to earn money for the future.

Resources and Tools

- **Activity Sheet 11.1:** Percentages (one for each student or pairs of students)
- **Activity Sheet 11.2:** What's My Rule? (one for each student or pair of students)
- **Activity Sheet 11.3:** Calculating Interest (one for each student or pair of students)
- **Resource 11.1:** Square Model (one projected for the class, or one copy for each student or group of students)
- Calculators (optional)

Preparation Prior to Lesson:

Sales Pitch: Display Resource 11.1 for the students or make enough copies for each student or groups of students.

Part 1: Prepare enough copies of Activity Sheet 11.1 for each student.

Part 2: Prepare enough copies of Activity Sheet 11.2 for each student.

Part 3: Prepare enough copies of Activity Sheet 11.3 for each student.

Children's Literature Option:

See the Literature for Financial Literacy list on p. 203 for ideas of children's books relevant to this lesson.

Key Language

interest: The money paid by a borrower as payment for borrowing money.

pattern: A repeated sequence.

percent: Part of 100.

rate: A quantity typically measured against another quantity.

Money Talks

Ask your students if they have a savings account in a bank. Chances are that those students who do have a savings account in a bank have a simple-interest savings account. This means that each month or year that you have your savings account, the bank deposits money into your account. This deposit is called **interest**.

Talk about the purpose of a savings account. Though most accounts are not currently paying high interest, having a savings account is still beneficial for keeping money in a safe and reliable institution. You still have access to your money whenever you need it, and the more money you save, the more you will have for future use!

Sales Pitch

Wouldn't it be great if you could get extra money in your bank account and you didn't have to do anything special? There are different types of bank accounts (places to hold your money) that receive different amounts of interest based on a **rate**. Explain that interest is calculated using percents. In the first part of the activity, students will explore how a **percent** is part of 100.

Display **Resource 11.1: Square Model** for the students, or make enough copies for each student or group of students. Ask, "How many different equations can you create to represent the number of squares in this figure?" Regardless of which model is used, students should arrive at the conclusion that there are 100 squares in the figure.

The Fine Print

Part 1: Percents

Explain to students that 100 is important when figuring out percentages. Each percent represents 1 out of 100. Have students work through **Activity Sheet 11.1: Percentages** to explore percents. Once the students are finished working, come back together as a class to explore their reasoning on how they completed the problems. Ask the following questions:

- What is the relationship that you see between whole numbers and percents?
- What do you think is the relationship between fractions and percents?
- How did you come to this conclusion?

Part 2: Patterns

In this activity, students will be examining a table of values and looking for **patterns**. Have students examine **Activity Sheet 11.2: What's My Rule?** individually, or in pairs. Explain to them that these tables are similar to what we would see on a bank account statement. The "in" can represent various figures, such as how much money you put into your account, and the "out" would be how much money you had after being paid interest. Or the "in" could be the number of years you have maintained the account and the "out" would be your balance on the account.

Standards for Mathematical Practice: Students develop MP.8 as they look for patterns for multiplying or dividing.

Emphasize the importance of identifying the pattern within the table and being able to model this pattern with a number expression or words. Students will also have the opportunity to apply their understanding by creating their own rule and describing the pattern. Students may find questions 6 and 7 challenging, and this is to be expected. Remind students how they calculated percentages on Activity Sheet 11.1.

Part 3: Calculating Interest

Explain to the students that now they will have a chance to deposit money into a hypothetical bank. Remind students of the previous conversation about how when you deposit money into an interest making account, your money will "grow." Review the vocabulary word interest. Give the students the following example: If you deposit $100 into your account that pays 2 percent interest, how much interest would you earn? Encourage students to talk through the amount that would be earned ($2.00).

Now have students do some research about different banks and interest rates. Share several findings about various accounts and their current interest rate and what this would mean for the $100 deposit. Remind students that a 2 percent interest rate, for example, means that the bank will give you $2 for every $100 dollars deposited into the account. Explain to students that most of these rates will change over time, depending on the economy. Ask students why they think there is variance in the rates instead of them all being the same percent.

Then have students work through **Activity Sheet 11.3: Calculating Interest** in pairs or as a class to help experience calculating simple interest. Some of your students might already have a bank account, and for those who don't, discuss the advantages of keeping your money in the bank, as opposed to a piggy bank in your room.

Closing the Deal

Remind students that they are modeling real-world mathematics as they use multiplication to calculate the amount of interest earned on their savings account.

To review, post an amount of money on the board, such as $44. Ask,

- How much money will I earn in one year if the interest rate is 10 percent?
- Does learning about interest change your views about saving money?

Close by having students write in their journal or say to a partner a pattern they noticed in the lesson that helped them solve the problems.

Getting Your Money's Worth

This activity can be modified for students who are ready for percents and decimals to the hundredths, or also for students who are struggling by allowing them to use a calculator to verify their answers.

References and Resources

Fidelity. "America's Savings Rate Improves, but Fidelity Study Finds More Than Half of Americans at Risk of Not Covering Essential Expenses in Retirement" (2016). https://www.fidelity.com/about-fidelity/individual-investing/americas-savings-rate-improves.

Jump$tart Coalition for Personal Financial Literacy. *National Standards in K–12 Personal Finance Education.* Fourth Edition. Second Printing. Washington, D.C.: Jump$tart Coalition for Personal Financial Literacy, 2017.

National Governors Association Center for Best Practices and Council of Chief State School Officers (NGA Center and CCSSO). *Common Core State Standards for Mathematics.* Washington, D.C.: NGA Center and CCSSO, 2010.

Activity Sheet 11.1

Percentages

1. One percent (1%) means 1 out of 100. How could you use the figure below to show 1 out of 100?

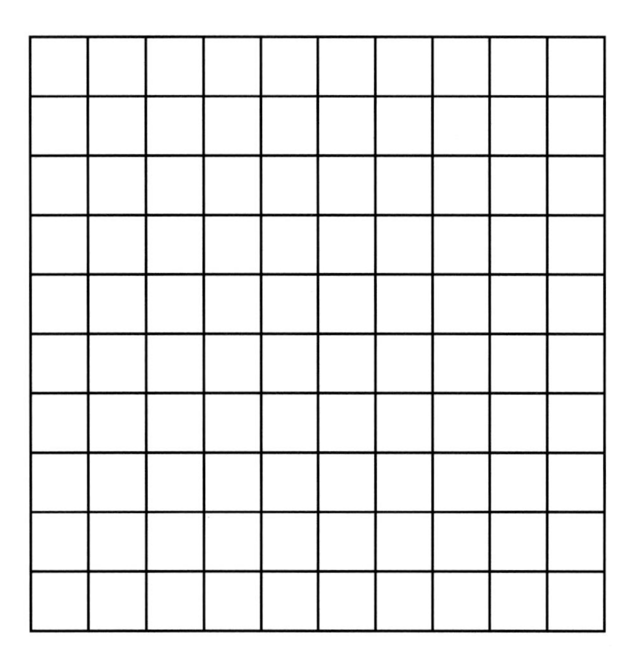

2. Fifty percent (50%) means 50 out of 100. How could you use the figure below to show 50 out of 100?

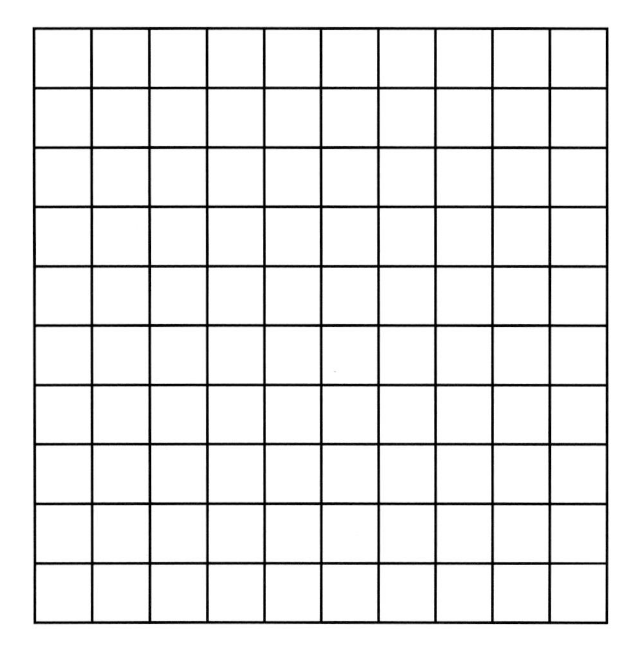

Activity Sheet 11.1—*page 2*

Percentages

3. Describe what percent of the figure below is shaded. How do you know?

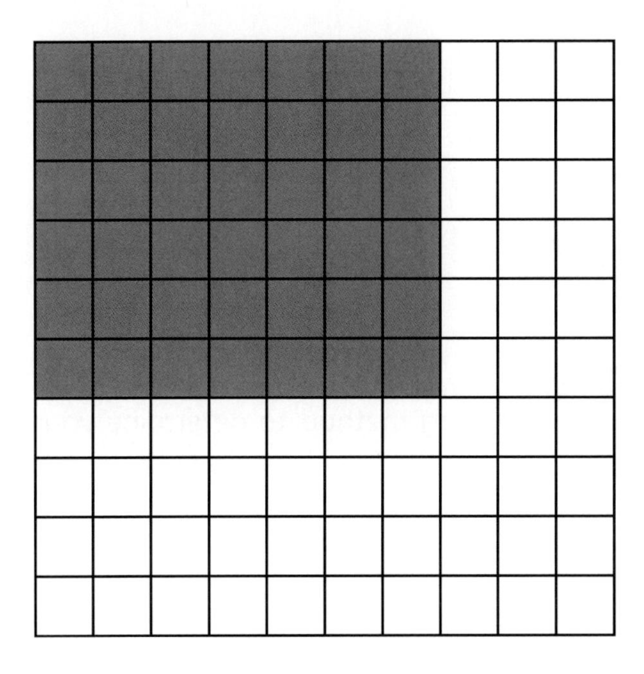

4. Use the figure below to model 20 percent (20%).

Activity Sheet 11.1—*page 3*

Percentages

5. Challenge: Use the accompanying figure to model 50 percent (50%). Explain your thinking in words.

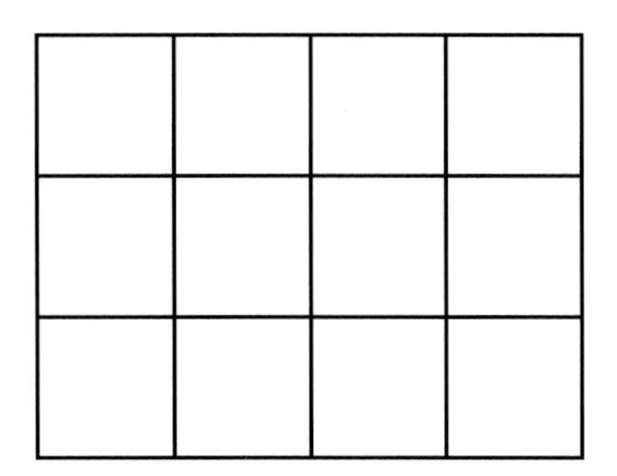

6. Let's go back to the first problem. If we have $100 and we want to represent each square as $1, what is 10 percent (10%) of $100? Use the square shading method to determine your answer.

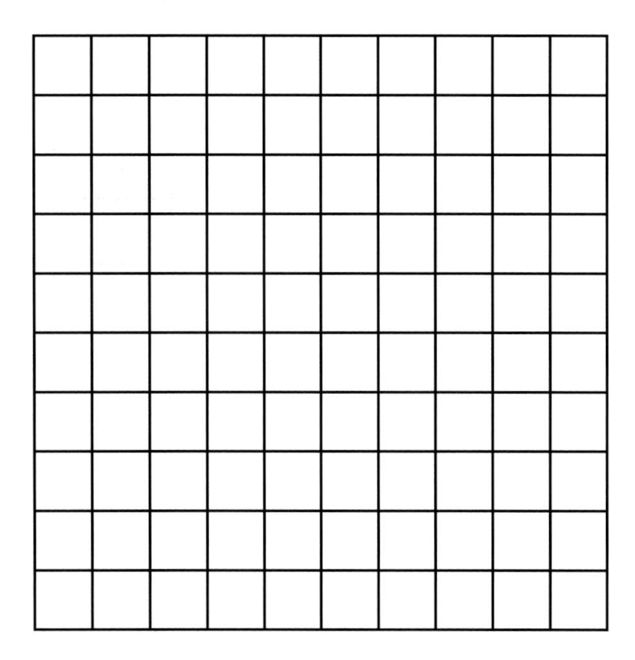

7. Many banks will give you interest on the money that you have in your bank account. If you have $100 in your account and the bank will pay you one percent (1%) interest, how much interest will the bank pay you?

Activity Sheet 11.2

What's My Rule?

Instructions: Use the following tables to determine the patterns for each problem.

1. Describe the pattern to go from the "in" to the "out."

In	Out
50	250
100	300
130	330
170	370

2. The rule is times 10. Find the numbers that complete the "out" column.

In	Out
2	
3	
5	

3. The rule is divide by 2. Find the numbers that complete the "in" column.

In	Out
	8
	6
	2

Activity Sheet 11.2—*page 2*

What's My Rule?

4. Describe the pattern to go from the "in" to the "out."

In	Out
2	16
3	24
7	56

5. Write your own rule and fill in the table.

In	Out

6. In this problem, you will compare the number of years that you've had your money in a bank account with the ending balance of your account at the end of each year. What is the amount changing by each year?

Years	Balance
1	$110
2	$120
3	$130
4	$140

7. In problem 6, you calculated the difference in the amounts each year. Another way to consider that problem is that you deposited $100 in your account and earned interest each year. If your account balance was $110 at the end of the first year, what percent interest did you earn to go from $100 to $110?

Activity Sheet 11.3

Calculating Interest

Instructions: In this activity, you will use the formula **Principle** × **Rate** × **Time** to calculate the amount of interest for a simple-interest savings account. **Principle** relates to the starting amount of money, **Rate** is the interest rate in percent, and **Time** is how long the money will remain in the account.

1. How do you convert 50 percent (50%) into decimal form?

2. How do you convert 10 percent (10%) into decimal form?

3. How do you convert one percent (1%) into decimal form?

4. If you deposit $100 (principal) into an account for 2 years (time) at an interest rate of 2% (2% = 0.02 rate), how much interest would you earn over those 2 years?

5. If you deposit $1000 (principal) into an account for 2 years (time) at an interest rate of 2% (2% = 0.02 rate), how much interest would you earn over those 2 years?

6. Use the amount of interest you calculated in question 5 to determine what your total balance would be at the end of 2 years. What is your balance (in dollars) at the end of 2 years?

Resource 11.1

Square Model

Write as many ways as you can think of to represent the number of individual squares in the figure below. An example would be $10 \times 10 = 100$. Model your thinking for each equation.

Chapter

How to Be Financially Responsible

The Bottom Line

Spending and saving money are two aspects of finance that many adults, let alone primary students, find challenging.

Students are provided with different scenarios of individuals who spend and save differently. They perform calculations involving fractions, while learning about spending and saving at the same time.

Standards Met in This Chapter

CCSSM Standards and Practices

3.NF.A.1: Understand a fraction $1/b$ as the quantity formed by 1 part when a whole is partitioned into b equal parts; understand a fraction a/b as the quantity formed by a parts of size $1/b$.

4.NF.B.3.a: Understand addition and subtraction of fractions as joining and separating parts referring to the same whole.

5.NF.B.3: Interpret a fraction as division of the numerator by the denominator ($a/b = a \div b$). Solve word problems involving division of whole numbers leading to answers in the form of a fraction or mixed numbers.

MP.2: Reason abstractly and quantitatively

MP.4: Model with mathematics.

MP.7: Look for and make use of structure.

(National Governors Association Center for Best Practices and Council of Chief State School Officers [NGA Center and CCSSO] 2010)

Jump$tart Standards

Saving and Spending: Apply strategies to monitor income and expenses, plan for spending and save for future goals.

Standard 1. Develop a plan for spending and saving.

Knowledge Statements:
Overall:

- Saving means choosing to set aside money for emergencies and future needs and goals.

Standard 1:

- Explain how saving money can improve financial well-being.
- Demonstrate how to allocate weekly income for spending, saving, and sharing goals.

(Jump$tart Coalition for Personal Financial Literacy 2017)

Future Value

Why is it challenging for many students to save money? No matter your age, it can be difficult to discern the difference between a want and a need. A lack of understanding on how to create a budget, and stick to it, can also generate problems. Students who begin to have a grasp on saving, spending, and budgeting will have a greater chance of finding themselves in better financial shape moving forward. This means less debt and better planning for the future, including saving for retirement.

Students will be using fractions to divide their money into saving, spending, and sharing jars. It is important for students to understand a fraction $1/b$ as the quantity formed by 1 part when a whole is partitioned into b equal parts. By gaining this foundation, students will have a better understanding of the composition of fractions and how to perform mathematical operations using fractional parts.

The following sections describe the mathematical and financial literacy concepts that this lesson will encompass, as well as the knowledge and experiences that elementary students may bring to the lesson and later take from it.

Mathematics

This lesson focuses on mathematical operations involving whole numbers and fractions. When working with numbers in the real world, many of those numbers will be fractions. Students should be able to explain that the unit fraction $1/b$ is the quantity formed by 1 part when a whole is partitioned into b equal parts (3.NF.A.1). For example, if I have five students whom I would like to split money between (b equal parts), each student will receive 1/5 of the money ($1/b$).

Older students will be able to conceptualize this notion and come to the conclusion that a fraction is division of the numerator by the denominator (5.NF.B.3). Students should also be able to explain that if two fractions represent the same whole and have the same denominator, their sum is found by adding the numerators together (4.NF.B.3.a). To achieve this level of understanding, students will need to reason abstractly and quantitatively (MP.2) and look for and make use of structure (MP.7).

Financial Literacy

In this activity, students will be able to make connections with characters in fictitious stories and see how they might react within a certain situation. Students need to have a chance to make mistakes with money when they are young and to learn from those mistakes before they encounter real financial trouble.

Student Knowledge and Experience

Students in grades 3 through 5 continue to develop their skills in performing operations with whole numbers and fractions. It is important for students to see the relationship between unit fractions and the whole number before they start to make calculations based strictly on memorization.

Students also might not have had experience maintaining a running balance for an account. This chapter emphasizes the importance of keeping accurate records and partitioning money as a means of saving for future goals.

Lesson Plan

Learning Targets

- Explain why it is important to save money for future needs.
- Maintain an accurate record of money saved.
- Calculate how to divide an amount of money into thirds.

Resources and Tools

- **Activity Sheet 12.1:** Savings Account Statement
- **Activity Sheet 12.2:** Save, Spend, Share Budget
- **Resource 12.1:** Scenario Cards
- **Resource 12.2:** Items to Purchase 1–2
- Die (one for each student or pair of students)
- *Sonja Meets Her Future Self* by the Women's Institute for Secure Retirement (2014) (or another book about saving for the future)

Preparation Prior to Lesson:

Sales Pitch: Access the video or obtain the book *Sonja Meets Her Future Self* by the Women's Institute for Secure Retirement.

Part 1: Prepare enough copies of Resource 12.1 for each student.

Part 2: Prepare enough copies of Activity Sheet 12.1 and Resource 12.2 as needed for each student. Obtain enough dice to provide one for each student or pair of students.

Children's Literature Option:

See the Literature for Financial Literacy list on p. 203 for additional relevant children's books.

Key Language

fraction: A numerical quantity that is not a whole number.

saving: Putting money away for the future.

sharing: Giving something of your own to others.

spending: Using money to get something that you want.

Did you know?
Ever wonder where the dollar sign ($) came from? The most widely accepted explanation, according to the Bureau of Engraving and Printing, goes back to the Spanish peso, which was the basic unit of value in colonial America during the late 1700s. Handwritten manuscripts from that time show that the peso—formally *peso de ocho reales*, or "piece of eight" in America—was abbreviated PS. It's believed that as time went on, the abbreviation was often written with the S on top of the P, somewhat like the $ symbol. The $ sign first appeared in print after 1800 and was widely used by the time the first U.S. paper dollar was issued in 1875 (History.com 2012).

Money Talks

Begin with a discussion by asking students if they know what it means to save money? Ask the following questions:

- Have you ever received money as a gift or from an allowance?
- If so, what do you do with your money? Do you spend it right away or do you save it?
- If you do save it, where do you keep your money so it doesn't get lost?

Sales Pitch

With your students, view the video or read the book *Sonja Meets Her Future Self* by the Women's Institute for Secure Retirement. (See the References and Resources section for information on how to do this.) Some questions you can ask include the following:

- If you could jump on a skateboard and travel in time, would you travel back in time or into the future? What would you want to see or visit?
- If you could **spend, save,** and **share** your money, what would you spend it on? With whom would you share your money?

Discuss with students that when people save money, they save for different reasons. Sometimes you might save money to buy something a few weeks away, or you might save money for years to be able to stop working when you want to. Ask students what might be some things that they would want to save for in order to purchase later.

Fine Print

Part 1: Scenario Cards

For this part of the activity, pair up your students. Each student will take a card from **Resource 12.1: Scenario Cards** and read it out loud to his or her partner. Once each pair has had the chance to share and discuss the two scenario cards, bring everyone back together as a large group. Ask your students the following questions:

- Which of these personalities, Spencer or Savannah, do you see as more like you? Why?
- What are some financial things that Spencer Spender should be worried about based on how he spends money?
- What are some things that both Savannah Saver and Spencer Spender can do to help their financial situation?
- How did you figure out how much money Savannah had to deposit, and how much candy Spencer could buy? Show your work and thinking.

Next, share scenario 3 on Resource 12.1 with your students. Have students model the mathematics by writing an equation to represent the scenario. Ask the following questions:

- Why do you think both Savannah and Spencer decided on saving half of their money?
- Do you agree with their decision? Why or why not?

Standards for Mathematical Practice: Students are developing MP.4 by modeling equations to represent Savannah and Spencer's financial decisions.

Part 2: On a Roll to Saving Money

For this activity, students will be paired or in small groups. Each group will draw one card from **Resource 12.2: Items to Purchase** to discover what they are saving their money to buy. They will then make a prediction of how many weeks it will take them to save that much money on **Activity Sheet 12.1: Savings Account Statement**. Each student will next roll one die to represent how much money they saved that week and record this amount on their savings account statement. (If you would like to make the game more difficult, use multiple die to represent multiple place values. For instance, use a die with only 0, 1, and 2 on it to represent the tens place and a regular die for the ones.) The goal is to accurately record and calculate your savings to see who can save enough money first to make the purchase.

Part 3: Partitioning the Money

This section gives the students an opportunity to use the ideasof spending, saving, and sharing to create a budget involving the unit **fraction** of 1/3. Students will be told an amount of money they have received. (Choose an amount that can be divided equally into thirds.) Explain to your students that the unit fraction 1/3 is the quantity formed by 1 part when a whole is partitioned into 3 equal parts.If we wanted to find out what 1/3 of a number is, we would divide that number by 3. For instance, Julio received $15.00 for his birthday. He would like to divide his money equally into three jars: save, spend, and share. He needs to figure out what 1/3 of $15.00 would be. To do that, he will need to divide the whole amount ($15.00) into three equal parts. Have the students model this with an equation 1/3 + 1/3 + 1/3 = 3/3. If the whole amount is $15.00, then $5.00 + $5.00 + $5.00 = $15.00.

Have the students record specified amounts on **Activity Sheet 12.2: Save, Spend, Share Budget** (an example is given at the top of the sheet). After students have had time to do a few lines, ask them to make some observations about their chart. Ask the following questions:

- Do you notice any patterns in your chart?
- What is the relationship between the save, spend, and share columns and the total column?
- Why do you think someone would want to organize their money into save, spend, and share jars?

Closing the Deal

Ask students which character, Spencer or Savannah, they feel portrays them the best. When they receive money, do they spend it right away, save it for later, or some combination of the two? Give students the scenario that they earned $30 cleaning up and recycling cans and bottles in their neighborhood. You would like them to divide the money into the save, spend, and share jars. Ask the following questions:

- If you need to divide your money into three jars equally, what fraction of your money will be in each jar?
- What equation would model this scenario?
- How much money will be in each jar?

If you have time, make a language arts connection by giving your students an opportunity to write their own short story. Explain to them that they can create fictional characters who make financial decisions, but they must include an operation with fractions in the story.

Getting Your Money's Worth

This lesson gives students the ability to begin working with fractions through the use of saving, spending, and sharing money. The activity can be modified by changing the fractional parts for each jar, or it can be geared toward higher-level students by asking students to multiply fractions. For example, make the savings jar worth 1/2 and the spending and sharing jars worth 1/4. You could also use story problems, such as this one: Savannah has some money and spent 2/5 of her money on clothes. She earned $5 the next day. Later, she lost 3/7 of her money, but she still had $20 left. How much money did Savannah have in the beginning before she spent some money on clothes? While this problem will pose a challenge to students, it will open the discussion to multiple strategies that can be used for problem solving.

References and Resources

History.com. "Where Did the Dollar Sign Come From?" (2012). http://www.history.com/news/ask-history/where-did-the-dollar-sign-come-from.

Jump$tart Coalition for Personal Financial Literacy. *National Standards in K–12 Personal Finance Education.* Fourth Edition. Second Printing. Washington, D.C.: Jump$tart Coalition for Personal Financial Literacy, 2017.

National Governors Association Center for Best Practices and Council of Chief State School Officers (NGA Center and CCSSO). *Common Core State Standards for Mathematics.* Washington, D.C.: NGA Center and CCSSO, 2010.

Women's Institute for Secure Retirement. *Sonja Meets Her Future Self.* (2015). A book version of this story can be downloaded at https://www.wiserwomen.org/images/imagefiles/sonya-meets-future-self-reduced.pdf; and a video version can be viewed at https://www.youtube.com/watch?v=i9Pbgu5unKg.

Activity Sheet 12.1

Savings Account Statement

What are you saving to purchase?

How much does it cost?

How many weeks do you think it will take you to save this much money?

Savings Account Statement: Create a table to represent the number of weeks, amount earned for that particular week, and a running total of how much has been saved in the account.

Week Number	Amount Earned This Week	Total Earned

How many weeks did it actually take you to earn the money to make your purchase?

Was this more, less, or the same as you predicted?

Activity Sheet 12.2

Save, Spend, Share Budget

Date	Amount Received ($)	Save ($)	Spend ($)	Share ($)
June 1, 2020	*15.00*	*5.00*	*5.00*	*5.00*
Total:	*15.00*	*5.00*	*5.00*	*5.00*
July 4, 2020	*24.00*	*8.00*	*8.00*	*8.00*
Total:	*39.00*	*13.00*	*13.00*	*13.00*
Total:				
Total:				
Total:				
Total:				
Total:				
Total:				
Total:				

Resource 12.1

Scenario Cards

Instructions for teacher: Copy enough scenario 1 and 2 cards for each of your students to have one scenario. Pair your students up, with each student in the pair having a different scenario. Scenario 3 will be done either in pairs, in small groups, or with the class as a whole.

Scenario 1

I'd like to introduce myself. My name is **Savannah Saver**. I am ten years old, and I earn money from helping my elderly neighbors with yard work and shoveling snow in the winter. I also just had a birthday. Many of my friends brought presents to my party, and a few of them gave me money. I received $5, $10, and $18 from my friends. I also helped rake leaves for one neighbor and earned $11.50. I don't have any idea what I want to do with my money, so I decided to save all of it. I always ask my parents to deposit all my money into my savings account in a bank. Can you help me figure out how much money I have to deposit?

Scenario 2

Hello! My name is **Spencer Spender**. I am nine years old, and I earn money from doing chores for my parents. Recently, I helped wash the dishes and I cleaned my room. My parents gave me $2.50 for washing the dishes and $1.30 for cleaning my room. I went to the store with my parents and wanted to buy some candy and a balloon. My parents asked me how much money I have to spend. Can you help me? How much money did I receive from doing my chores? I don't want to save any of my money, and I want to spend it all on candy and a balloon. The balloon I want to buy is $0.70 and I want candy that costs $0.50 each. How much candy can I buy if I also want to buy one balloon?

Scenario 3

In this story, both Savannah and Spencer will be working together to help each other make decisions about their money. Savannah has a total of $44.50 in her savings account, and Spencer has $3.80 before he goes to the store with his parents. Savannah and Spencer talk about some things they might want to do with their money. Spencer wants to buy candy and a balloon. Savannah decides she wants some new clothes, and finds out that the clothes will cost a total of $60. After talking, they think it is best if they spend some of their money and save some of their money. Savannah and Spencer have agreed that spending 1/2 and saving 1/2 of their money will be what is best currently, and for the future. If they both want to save 1/2 of their money, how much will each of them be saving?

Resource 12.2

Items to Purchase 1

Scrapbook Kit $20.00	Tote Bag $11.00	Book $10.00
Necklace $15.00	Baseball $12.00	Water Bottle $6.00
Markers $13.00	Journal $14.00	Hair Chalk $8.00
Trading Cards $10.00	Puzzle $9.00	Volleyball $15.00
Model Airplane $16.00	Chess Set $17.00	Yo-Yo $7.00

Resource 12.2

Items to Purchase 2

Exercise Tracker $50.00	Microscope $49.00	Magic Tricks Kit $37.00
Robot $45.00	Trampoline $35.00	Stuffed Animal $15.00
Basketball $22.00	Bicycle $54.00	Music Player $32.00
Dart Board $19.00	Painting Set $23.00	Tennis Racket $34.00
Modeling Clay $20.00	Headphones $27.00	Beanbag Chair $42.00

Chapter

Boardwalk Budget

The Bottom Line

It is beneficial to budget your money before making financial decisions.

Students use their mathematics skills during a trip to the boardwalk, where they will make decisions based on a financial budget.

Standards Met in This Chapter

CCSSM Standards and Practices

3.NBT.A.2: Fluently add and subtract within 1000 using strategies and algorithms based on place value, properties of operations, and/or the relationship between addition and subtraction.

4.NBT.B.4: Fluently add and subtract multi-digit whole numbers using the standard algorithm.

4.NF.C.6: Use decimal notation for fractions with denominators 10 or 100. *For example, rewrite 0.62 as 62/100; describe a length as 0.62 meters; locate 0.62 on a number line diagram.*

5.NBT.A.3: Read, write, and compare decimals to thousandths.

5.NBT.B.7: Add, subtract, multiply, and divide decimals to hundredths, using concrete models or drawings and strategies based on place value, properties of operations, and/or the relationship between addition and subtraction; relate the strategy to a written method and explain the reasoning used

MP.6: Attend to precision.

(National Governors Association Center for Best Practices and Council of Chief State School Officers [NGA Center and CCSSO] 2010)

Jump$tart Standards

Spending and Saving: Apply strategies to monitor income and expenses, plan for spending and save for future goals.

Standard 3. Describe how to use different payment methods.

Knowledge Statements:
Overall:

- A spending plan (budget) is a guide to help people balance money coming in (income) and money going out (expenses).

Standard 3:

- Verify the total cost of a purchase that includes multiple items.
- Calculate the amount of change to be returned when the payment amount is greater than the purchase price.

(Jump$tart Coalition for Personal Financial Literacy 2017)

Future Value

As they grow older, students will be faced with more and more choices of how and when to spend their money. Practicing how to create a budget and managing to stick to it is a major milestone in learning not to overspend or live beyond one's means. Students will also need to be aware of the cost of items and be able to calculate a running total to know how much they have spent and on what. Using mathematics skills to plan and adhere to a budget is a lifelong lesson in financial management and promotes positive spending habits.

The following sections describe the mathematical and financial literacy concepts that this lesson will encompass, as well as the knowledge and experiences that elementary students may bring to the lesson and take away from it.

Mathematics

This chapter focuses on the reinforcement of operational calculations with precision (MP.6) when using whole numbers and decimals (4.NF.C.6; 5.NBT.A.3). Students will be adding and subtracting multiple numbers (3.NBT.A.2; 4.NBT.B.4; 5.NBT.B.7) and will need to be cognizant of place value and decimal location. They will also use these skills while making change and performing mental mathematics as they estimate how much they are spending and whether this amount will stay within their budget.

Financial Literacy

One of the main concepts that students must understand on their mission toward financial literacy is budgeting. Financially literate students need to be able to create a budget, perform calculations within that budget, and realize that budgets are subject to change. People frequently find themselves in situations where they have overspent or were unprepared for an unforeseen circumstance. Budgeting and saving will help offset these instances and provide the stability needed to make it through difficult times.

In this lesson, students will also be practicing the skill of making change. Although the use of bills and coins has subsided in the wake of debit and credit cards, there are still many instances where it will be necessary for students to be able to perform the correct calculation to determine the accurate change due to a customer.

This activity will provide students with the opportunity to create a budget—and then use that budget to become an entrepreneur! Students will be divided into two groups, buyers and sellers, and they will create a situation similar to one you might find at a boardwalk in the summer. Students will create their own advertisements to help increase exposure for their business.

Student Knowledge and Experience

Students in grades 3 through 5 have varying experience in performing operations with whole numbers and decimals. Even if students are not yet up to decimal calculations, they will benefit from the amount of whole-number manipulation that occurs in this chapter. Students can also apply their knowledge of rounding and estimation to help them create and manage a budget.

Some students may have previous experience with spending money, such as while purchasing candy. Make connections to these past experiences and build on them with the notion that money is not limitless and we sometimes have to make difficult decisions when determining how to spend our money. These notions can lead to further discussion on how we don't have to spend money immediately when we receive it, but rather we can save up for a future purchase or something that might be more expensive.

Lesson Plan

Learning Targets

- Students will perform calculations using whole numbers and decimals.
- Students will create a budget for a vacation purchase.
- Students will act as both a buyer and seller and make financial decisions based on their resources.

Resources and Tools

- **Resource 13.1:** Beach Items 1–2 (make enough copies so that several students may purchase the items from the stations)
- *Our Vacation Budget: Working with Decimals* by Andrew Einspruch (2013) (or another book about budgeting)
- Calculators (optional)

Preparation Prior to Lesson:

Sales Pitch: Obtain a copy of *Our Vacation Budget* by Andrew Einspruch.

Part 1: None.

Part 2: Prepare enough copies of Resource 13.1 so that each seller can have a stack of their items at their station.

Children's Literature Option:

See the Literature for Financial Literacy list on p. 203 for other relevant children's books.

Key Language

budget: To allow for a set amount of money.

buyer: The person making the purchase.

seller: The person who has something to sell.

Money Talks

Begin by asking students if they have ever been on vacation. Allow them to share a few stories. Ask them if they ever considered the expense of that vacation and how long their parents or guardians might have had to save before they were ready to take the trip. Explain to students that sometimes a lot of planning happens before the vacation or other significant event that requires thought and budgeting. Brainstorm with the students what budgeting means to them. After students have shared their ideas, explain to them that sometimes people make **budgets** in order to know what their money is being spent on and to make sure they do not overspend and end up without any money available. People also make a budget for how much they want to spend on a particular item or event so they can plan to save and do not overspend when it is time to make a purchase.

Sales Pitch

Read *Our Vacation Budget* by Andrew Einspruch with the class. The first few pages of the book talk about creating a budget for going on vacation. Early on, a section asks students to create a table using place value to organize some of the expenses of a vacation. Talk with students about the benefits of creating a budget and how this can apply to their current lives. Explain to them that many times we use rounding to estimate how much money we think we will need.

The Fine Print

Part 1: Budgeting

Tell students to imagine that they are going to the beach on vacation in a few weeks. Ask the following question:

- What are some items you might want to buy when you are on vacation? (Ideas might include a bathing suit, beach towel, beach umbrella, beach chair, beach ball, sunglasses, sunscreen, or boogie board.) Record the brainstormed list on the board.

Next ask students this question:

- What do you think these items might cost? (If students are unsure, do a quick search on the Internet for current prices and record this value in whole numbers or with a decimal to the hundredths place if appropriate.

Explain to students that their grandmother gave them $50.00 to spend on vacation. Instruct students to create a list of *at least* three items that they intend to purchase. Along with the list of objects, students should include the cost for each. Once students have their lists complete, ask them to write an expression that would model the total estimated cost for their vacation items. Students need to make sure they stay within their $50.00 budget.

Now tell your students that you just heard about a hurricane that made landfall at your vacation spot! By the time you are scheduled for your vacation, things will be cleaned up and ready to go, but due to the hurricane, the prices have doubled for everything on each student's list. Have your students recalculate the new total cost of their items. Ask students the following questions:

- If this was a real scenario, would you have enough money for everything you want on vacation?

- If not, what are some things you could do to close the gap between what you want, and how much money you have to spend? (Ideas might include doing work for money or saving their allowance.)

Did you know?
The oldest and also longest boardwalk in the United States is in Atlantic City. According to National Geographic, the first wooden planks were laid in 1870 to help keep the sand on the beach from entering the train and hotel lobbies. The boardwalk spans over 5.5 miles long! (Loftus 2010).

Part 2: Buyers & Sellers

This part of the activity can be as creative as you choose. Students can dress up like they are going to the beach and really have fun with it! Divide your class into **buyers** and **sellers**. This can be done as a whole class, or set up as small groups. The buyers will need a variety of currency to purchase items (around $50.00, so use play money). The sellers will also need currency to make appropriate change. Sellers will need to create a sign advertising their products and the cost of the products. Sellers then need to set up their stations with their items. Make enough copies of **Resource 13.1: Beach Items** so that each seller can have a stack of items at his or her station with price signs (similar to what you might find on a boardwalk). If students are not familiar with a boardwalk, you can show them an online article (Firpo-Cappiello 2013) with nineteen examples.

Buyers will then visit the different "shops" to purchase items. You might suggest to the buyers that they visit the shops first, without buying, and make a budget for how they plan to spend their money. They can take notes in their journal so that they will have a reference when they are ready to make their final financial decisions. As students "purchase" items, both the buyer and seller must record the transaction in their journal. Each will write the initial price of the item, how much money was given to pay for the item, and how much change was received, if applicable. It is important for students to attend to precision when making change in order to not shortchange their buyer. Students can continue to make purchases until they are out of money. Sellers should calculate how many items they sold and the total amount of money they made that day. Buyers and sellers can then switch positions to see a different aspect of the economic experience.

Closing the Deal

After both groups have had the opportunity to shop and calculate their earnings, bring the class back together to discuss what they have learned. Ask students,

- Were you able to purchase everything you wanted?
- Did anyone go over budget, or borrow money from a friend?
- Did you find it harder to be the buyer or seller? Why?
- Was it difficult to keep track of the calculations and make change?
- If you didn't have to spend any of your money, would you?
- Why do you think it is important to make a budget or plan before visiting a boardwalk or other shopping avenue?

Getting Your Money's Worth

Boardwalk Budget is a fun activity that lets your students' personalities shine through, along with their creative entrepreneurial spirit! These activities can be adapted for many grades depending on ability level. Consider extending the activity to have markdowns and sales where students will have to calculate how much of a discount they will receive. You could also incorporate rounding skills by having the students estimate their budget first (3.NBT.A.1; 4.NBT.A.3; 5.NBT.A.4). Giving students the opportunity to practice budgeting and thinking before spending will be a lifelong lesson that they can take with them as they become adults.

References and Resources

Einspruch, Andrew. *Our Vacation Budget: Working with Decimals*. Huntington Beach, Calif.: Teacher Created Materials, 2008.

Firpo-Cappiello, Robert. "19 of the Best Boardwalks in America." CNN Travel (2013). http://www.cnn.com/travel/article/america-best-boardwalks/index.html.

Jump$tart Coalition for Personal Financial Literacy. *National Standards in K–12 Personal Finance Education*. Fourth Edition. Second Printing. Washington, D.C.: Jump$tart Coalition for Personal Financial Literacy, 2017.

Loftus, Margaret. "Top 10 U.S. Boardwalks." National Geographic.com (2010). http://www.nationalgeographic.com/travel/destinations/north-america/united-states/us-boardwalks/.

National Governors Association Center for Best Practices and Council of Chief State School Officers (NGA Center and CCSSO). *Common Core State Standards for Mathematics*. Washington, D.C.: NGA Center and CCSSO, 2010.

Resource 13.1

Beach Items 1

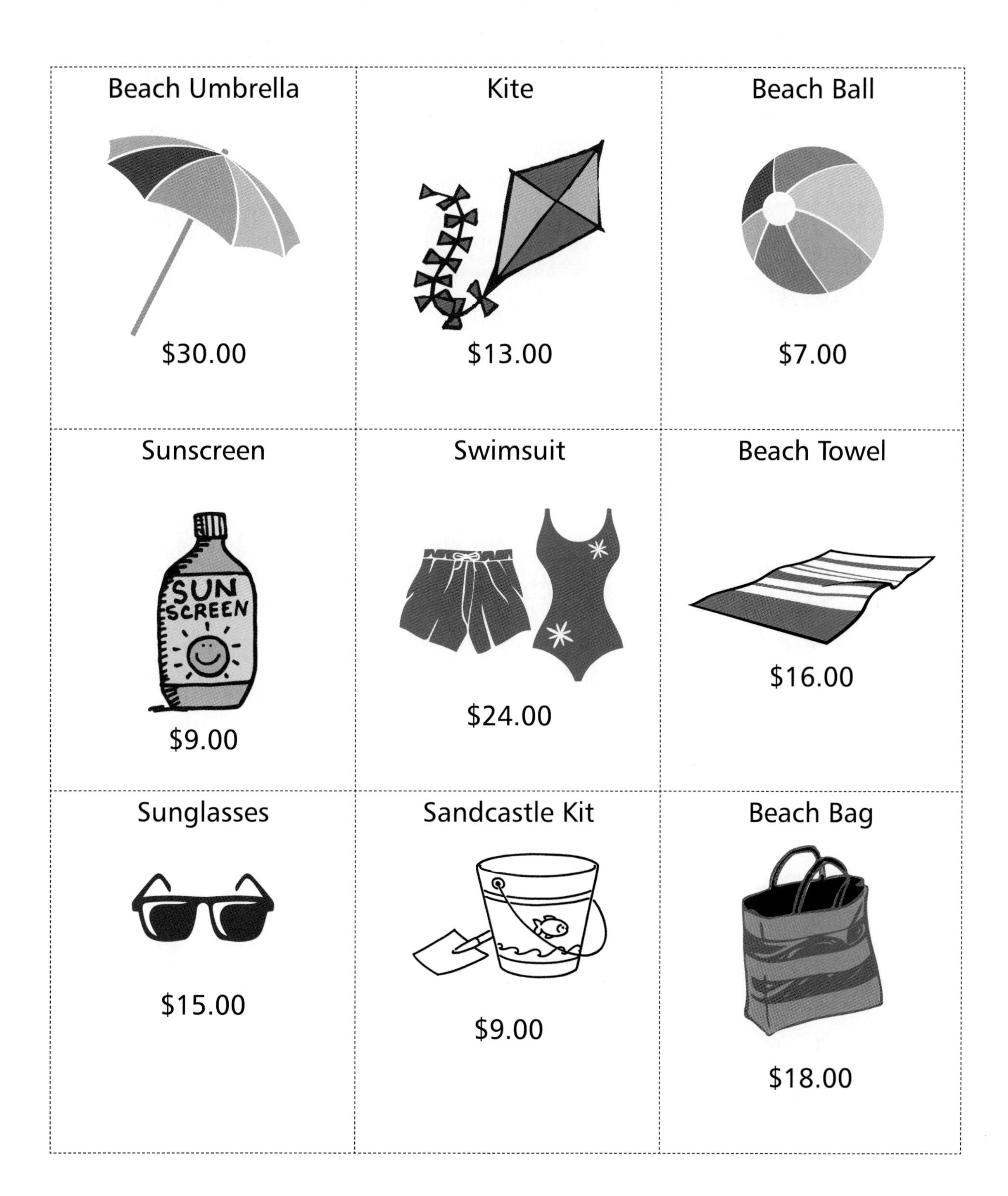

Resource 13.1

Beach Items 2

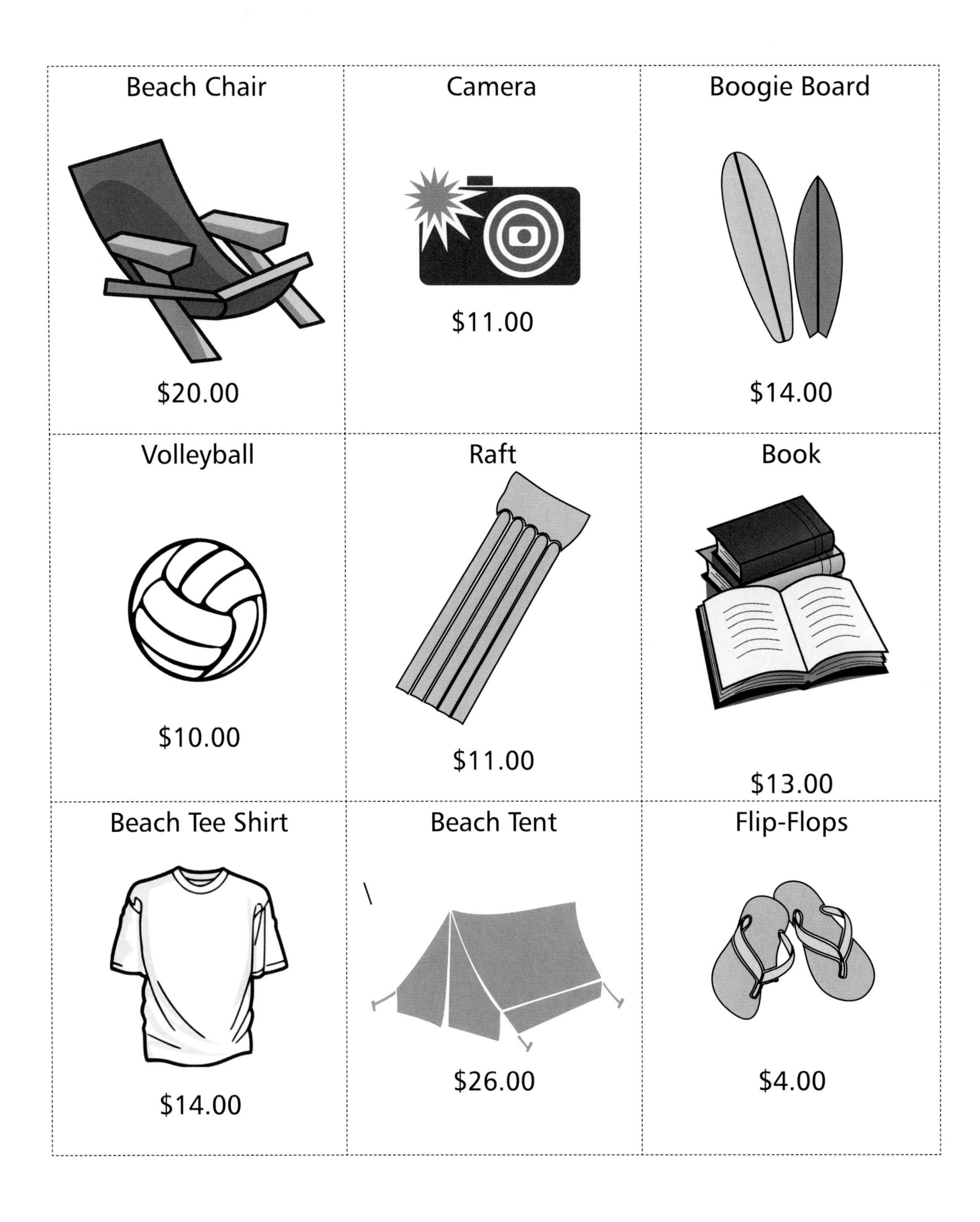

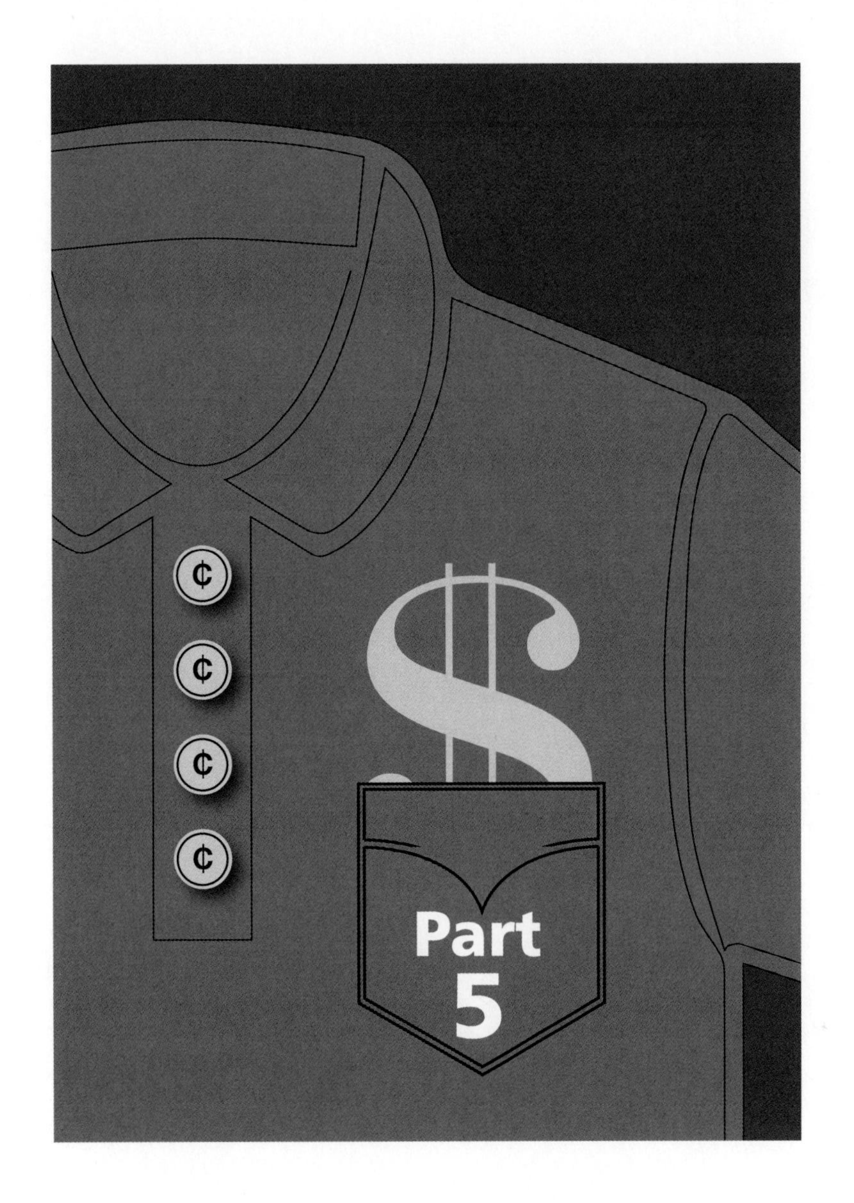

Measurement and Data

Exploring Jobs and Careers through Data

The Bottom Line

You can do some jobs when you are young and some when you get older; the jobs you choose to do are based on your interests and your needs.

Students experience the four-step statistics process connected to jobs they can do now and jobs or careers they might like to have when they get older. Bar graphs and pictographs are used to represent and analyze the data.

Standards Met in This Chapter

CCSSM Standards and Practices

1.MD.C.4: Organize, represent, and interpret data with up to three categories; ask and answer questions about the total number of data points, how many in each category, and how many more or less are in one category than in another.

2.MD.D.10: Draw a picture graph and a bar graph (with single-unit scale) to represent a data set with up to four categories. Solve simple put-together, take-apart, and compare problems using information presented in a bar graph.

3.MD.B.3: Draw a scaled picture graph and a scaled bar graph to represent a data set with several categories. Solve one- and two-step "how many more" and "how many less" problems using information presented in scaled bar graphs. *For example, draw a bar graph in which each square in the bar graph might represent 5 pets.*

MP.1: Make sense of problems and persevere in solving them.

MP.6: Attend to precision.

(National Governors Association Center for Best Practices and Council of Chief State School Officers [NGA Center and CCSSO] 2010)

Jump$tart Standards

Employment and Income: Use a career plan to develop personal income potential.

Standard 1. Explore job and career options.

Standard 2. Compare sources of personal income and compensation.

Knowledge Statements:

Overall:

- People have jobs that pay money.

Standard 1:

- Give an example of how an individual's interests, knowledge and abilities can affect job and career choice.
- Give examples of careers related to a personal interest.

Standard 2:

- Discuss special household tasks that might be suitable opportunities to earn money.

(Jump$tart Coalition for Personal Financial Literacy 2017)

Future Value

Of interest to all ages is finding a job you *can* do and that *needs to be done*. For young children, that is likely a household job. As they get older, it becomes jobs within the community, such as mowing lawns, babysitting, dog walking, or other such tasks. Choices of job opportunities grow to include working for businesses and eventually pursuing a career. Starting young by reflecting on opportunities and selecting those that align with your interests and abilities helps students begin to see their potential and the value of increasing their knowledge and skills to expand their money-earning opportunities.

The following sections describe the mathematical and financial literacy concepts that this lesson will encompass, as well as the knowledge and experiences that elementary students may bring to the lesson and take from it on its completion.

Mathematics

Data is a great "fit" for financial literacy lessons because both data and lessons are best when connected to the real interests and experiences of students. Doing statistics involves a four-step process (Franklin et al. 2005):

1. Formulating questions
2. Collecting data
3. Analyzing data (including the creation of graphs)
4. Interpreting results.

This process focuses on making sense of the situation and looking back and therefore is a strong connection to MP.1. Not all questions are created equally! Statistical questions anticipate variability in data related to the question. A question such as "What kind of jobs do the friends in our class like?" is a statistical question because we expect a variety of answers. A key to teaching statistics is to make sure the question is central to the lesson. When asking questions that focus on mathematics (e.g., "How many more students picked ___ than ___?") or technical components (e.g., "What labels do we need to include on our graph?"), make sure to also ask questions to provide insights into the question posed (e.g., "What does the graph tell us about our question?) (1.MD.C.4; 2.MD.D.10; 3.MD.B.3). In this lesson, students focus on the language involved in writing a statistical question (MP.6).

Financial Literacy

In this chapter, we will refer to a "job" as working to earn money at a position that could fall within the scope of a career. A second grader might do jobs as part of being a student or a family member, or they might do jobs in order to earn money. Hamm (2008) defines a job as "simply something you do to earn money. Career advancement is not something you're interested in there and the work often doesn't interest you at all. In five years, you'll likely not be doing anything like your current job." A career, on the other hand, is "a series of connected employment opportunities, where you build up skills at earlier employment opportunities to move you into higher paying and higher prestige employment opportunities later on. In five years, you're planning to be doing something very similar to what you're doing now, but hopefully with more income and more interesting problems to tackle" (para. 2–3).

A career is something young students will do when they are older, but they can begin thinking about their interests now so that they consider the wide array of possibilities and find a career to match their skills and/or interests in the future.

Student Knowledge and Experience

This lesson builds on students' prior knowledge of a job. They might know other terms for jobs: chores, responsibilities, work, and so on. They also are likely to know of many careers. With these ideas in their lived experiences, they can engage in meaningful discussions around what makes something a "job" versus a "hobby," or what makes something a "job" versus a "career." While young students are not likely aware of the monetary influences on career selection, they could be thinking of various reasons why a person might select a particular job or career (and why they might not).

Jobs and careers are subject to a lot of biases. This lesson is a great opportunity to make sure *every* child can see themselves in the careers that are being discussed. The Sales Pitch provides an opportunity to open up this dialogue.

Mathematically, students may have seen graphs and gathered data within the class on various topics. Bar graphs are the most concrete of the graph types in this lesson. Pictographs (also called picture graphs) may seem more concrete, but they can be more challenging to young students because the picture may represent different quantities. Because of this notion of a scale, picture graphs are a great connection to skip-counting (grades 1 and 2) and multiplication (grade 3).

Lesson Plan

Learning Targets

- Design questions that can be answered by gathering data.
- Organize and interpret data to ask and answer questions related to careers.
- Describe characteristics of jobs and careers.

Resources and Tools

- **Activity Sheet 14.1:** Picturing Jobs and Careers
- **Activity Sheet 14.2:** Jobs for Now and Jobs for Later
- Scissors (one pair per student)
- Glue sticks (one stick per student, or they can share)
- Graph paper (optional)
- For pictographs: cutouts of clip art images that look like the jobs (five to ten per job, per group)
- *Oh, the Places You'll Go!* by Dr. Seuss (1990) (or another book relating to job and career options)

Preparation Prior to Lesson

Part 1: Copy Activity Sheet 14.1 for each student or pair of students.

Part 2: Provide graph paper (optional) for each student or pair of students.

Children's Literature Options

Oh, the Places You'll Go! by Dr. Seuss; *Mighty Mommies and Their Amazing Jobs: A STEM Career Book for Kids* and/or *My Daddy, the Amazing Nurse!: A Rhyming Career Exploration Book for Children and Others* by Donald Jacobsen (author) and Graham Evans (illustrator).

See the Literature for Financial Literacy list on p. 203 for additional ideas.

Key Language

bar graph: A graph with horizontal or vertical bars to represent value.

career: An occupation that a person usually has for a significant time in their life and includes opportunities for advancement.

data: Pieces of information.

job: Work for which a person is paid.

pictograph (or picture graph): A graph in which pictures are used to represent the data; The picture can represent one piece of data or it can represent a designated quantity.

survey: A question or questions that are asked to a population, such as a classroom, to learn about the population.

Money Talks

Ask children to think of one thing they like to do—a sport, hobby, or activity. Next, ask students what they might like to do as a **job**. Ask questions such as the following:

- What is a job?
- How might a job be like a hobby or activity? How are they different?
- What might you need to know or do in order to get a job?

Talk with students about their "job" as a learner. Though they might not be paid with money, students do have responsibilities and other expectations that are just like paying jobs. Ask the following:

- What are some expectations I have for you in your work?
- What are some aspects you like about your job?
- What are some things you would like to change about your job?

Discussing students' interests will help them start thinking about careers that match their interests and strengths and about ways they might improve to become better learners.

Did you know?
The most common job in Florida, Massachusetts, Rhode Island, and West Virginia in 2014 was primary school teacher! (IPUMS-CPS/ University of Minnesota, as cited by Quoctrung Bui 2015).

Sales Pitch

Read the book *Oh, the Places You'll Go* by Dr. Seuss. Ask students to share ideas of different **careers** (a job they might want to do when they grow up). Distribute the career illustrations cards from **Activity Sheet 14.1: Picturing Jobs and Careers**. You can cut this page in half, or have students pick two different jobs or careers. Ask students to pick one (or two) careers that sound interesting to them and draw a picture of what a person in that career might *do* on a regular basis, and then list what their responsibilities might be.

The Fine Print

Part 1: Exploring Jobs We Can Do

Revisit the idea of a job. Beyond being a student, ask your students to share other jobs they can do *now.* Invite them to share their ideas with a partner and listen for new ideas. Distribute **Activity Sheet 14.2: Jobs for Now and Jobs for Later**. Ask students to cut out the squares with different jobs and to sort them and glue them in the space provided (Jobs I Can Do Now or Jobs I Can Do When I Get Older). Alternatively, you can post this list of jobs and ask students to identify the ones they can do while they are still in elementary school.

Formulating Questions. To begin the statistical process, ask students, "What questions do you have about our class related to jobs we can do now?" Listen to ideas and record them on the board. Then ask, "Which of these questions can be answered by asking all of our friends in this class?" Select a question from the list that is a statistical question. Examples include the following:

- What are our friends' favorite jobs they can do now? (This question is asked on Activity Sheet 14.2.)
- What jobs (chores) do our classmates do in their own home?
- What jobs have our classmates done, not including what they do in their home?
- What jobs do our classmates *least* want to do?

If working with younger students, you can create a list of three categories. For older students, you can have a few more categories than that.

Data Collection. Next ask students to talk to a partner and decide how they will gather this information from every person in the class. Allow discussion time and then listen to ideas and select a **data** collection strategy. (You can do a simple hand-raise, or you can give each pair of students the task of figuring out how to gather data on their own.) Explain to students the components of a **survey**.

If students have gathered data on their own, discuss how many students they have for each category. When answers are different, ask why that might be (e.g., one group didn't ask the same number of people, a person changed his or her mind, etc.)

Standards for Mathematical Practice: Students are able to model with mathematics by using graphs to analyze information and draw conclusions based on the data (MP.6).

Data Analysis. For creating a **bar graph**, you can do one quickly as a whole class by having students post stickers or sticky notes above the options for the survey (e.g,, favorite jobs). For a deeper focus on creating graphs, distribute graph paper and discuss with students how they can construct a graph and how they can represent the data they collected on their graph.

If you are creating a **pictograph** for the first time, have the picture (clip art) represent one person. If students are familiar with pictographs, have them create a scale, such as one picture represents two classmates.

Once graph(s) are completed, ask students to analyze what they see. Ask the following questions:

- What do you notice about this graph?
- How many classmates picked ____________?
- How many more picked ____________ over ____________?
- How many did not pick ____________?

Standards for Mathematical Practice: Students are making sense of problems through interpreting data by recognizing trends and constructing predictions based on a graph (MP.1).

Interpreting Results. Ask students to recall what question they were exploring. Then ask them to share one insight related to the question they asked. Listen to many ideas. Ask questions that go beyond the graph, such as the following:

- What did we *not* learn from this statistical question?
- Would this graph look the same if we surveyed [another class? another grade? middle school students?]
- Based on your data, what are you now wondering that might be answered by posing another question that can be answered by surveying our classmates again?

Part 2: Careers for When We Get Older

Read one of the other books suggested in the Children's Literature Option section above, or share information about popular careers in your area. If time allows, ask students to engage in a discussion with their family members about their careers and what they like or don't like about them. Repeat the four-step process described in the Mathematics section above, but this time with a focus on Careers for When We Get Older.

Closing the Deal

Ask students to write down one sentence telling a big idea they learned about jobs and careers. Allow different students to read their sentence. Next, ask students to think about what they did in order to learn about their classmates' interests in jobs and careers. Revisit the four-step process. Close by asking students to write or say a new question they wonder about that they think can be answered in a survey.

Getting Your Money's Worth

This activity can readily be modified for any age, simply by adapting the statistical process and the types of graphs used to represent the data. Beyond careers, students can explore other questions related to financial literacy (and questions beyond financial literacy, of course). In fact, the statistical process can be a weekly routine wherein a question is generated one day, data gathered the next day or two, graphs completed the following day, and data interpretation completed at the end of the week.

References and Resources

Dr. Seuss (Theodore Geisel). *Oh, the Places You'll Go.* New York: Random House Children's Books, 1988.

Franklin, Christine, Gary Kader, Denise Mewborn, Jerry Moreno, Roxy Peck, Mike Perry, and Richard Scheaffer. *Guidelines for Assessment and Instruction in Statistics Education (GAISE) Report: A Pre-K–12 Curriculum Framework.* Alexandria, Va.: American Statistical Association, 2005

Hamm, Trent. "The Difference between a Job and a Career." The Simple Dollar (2008). https://www.thesimpledollar.com/the-difference-between-a-job-and-a-career/.

International Labour Organization. (n.d.). "What Is a Minimum Wage." http://www.ilo.org/global/topics/wages/minimum-wages/definition/WCMS_439072/lang--en/index.htm

Jacobsen, Donald. *Mighty Mommies and Their Amazing Jobs: A STEM Career Book for Kids.* Independently published, 2017.

_____. *My Daddy, the Amazing Nurse!: A Rhyming Career Exploration Book for Children and Others.* Independently published, 2016.

Jenkins, Emily, and G. Brian Karas. *Lemonade in Winter: A Book about Two Kids Counting Money.* New York: Random House, 2012.

Jump$tart Coalition for Personal Financial Literacy. *National Standards in K–12 Personal Finance Education.* Fourth Edition. Second Printing. Washington, D.C.: Jump$tart Coalition for Personal Financial Literacy, 2017.

National Governors Association Center for Best Practices and Council of Chief State School Officers (NGA Center and CCSSO). *Common Core State Standards for Mathematics.* Washington, D.C.: NGA Center and CCSSO, 2010.

Quoctrung Bui. "Map: The Most Common Job in Every State." Planet Money/NPR (2015). https://www.npr.org/sections/money/2015/02/05/382664837/map-the-most-common-job-in-every-state.

Smith, Jane. "5 Most Important Aspects of a Job." Work – Chron.com (2018). http://work.chron.com/5-important-aspects-job-6301.html.

United States Department of Labor. "History of Federal Minimum Wage Rates under the Fair Labor Standards Act, 1938–2009" (n.d.). https://www.dol.gov/whd/minwage/chart.htm.

Activity Sheet 14.1

Picturing Jobs and Careers

Job or Career: ______________________

Draw a picture of someone doing this job or career.

List responsibilities or skills for a person in this job or career:

____

____

____

Job or Career: ______________________

Draw a picture of someone doing this job or career.

List responsibilities or skills for a person in this job or career:

____

____

____

Activity Sheet 14.2

Jobs for Now and Jobs for Later

Cut out these cards and glue or tape them in the box below where you think they belong.

Walk dogs	Be a veterinarian	Sell lemonade in your front yard	Teach 3rd grade
Work at the grocery store	Be an architect	Sell lawnmowers	Feed neighbors' pets when they are not at home
Be a flight attendant	Deliver newspapers	Clean up a yard	Drive a truck

Jobs I Can Do Now	Jobs I Can Do When I Get Older

Select ONE of the jobs you can do now as your favorite idea for earning some money. Write the job here:

____

Chapter

Keeping Private Information Private

The Bottom Line

We all have identifiable information that is either meant to be shared or meant to be kept private.

Students discuss and explore what information is meant to be kept to themselves, shared with a parent, or shared with friends. Students will also discover strategies and ways to keep private information private.

Standards Met in This Chapter

CCSSM Standards and Practices

K.MD.B.3: Classify objects into given categories; count the numbers of objects in each category and sort the categories by count.

1.MD.C.4: Organize, represent, and interpret data with up to three categories; ask and answer questions about the total number of data points, how many in each category, and how many more or less are in one category than in another.

MP.3: Construct viable arguments and critique the reasoning of others.

MP.5: Use appropriate tools strategically.

(National Governors Association Center for Best Practices and Council of Chief State School Officers [NGA Center and CCSSO] 2010)

Jump$tart Standards

Financial Decision Making: Apply reliable information and systematic decision making to personal financial decisions.

Standard 7. Control personal information.

Knowledge Statements:

Standard 7:

- Explain the reasons for not sharing personal information with strangers.
- List types of personal information that should not be disclosed to others in person or online.
- Apply strategies to protect personal information.

(Jump$tart Coalition for Personal Financial Literacy 2017)

Future Value

Each one of us has an identity that distinguishes us from other individuals. This identity includes physical characteristics, fingerprints, DNA, Social Security number, student ID, and other distinct personal information. Identity belongs to one's self, and each person has the option to share this information with others. Sometimes this disclosure is necessary, as when applying for a job, or using a Social Security number to open a bank account. Other times, this information should be kept private. In this chapter, we will explore when it is appropriate to keep information to one's self, share it with parents, or share it with friends. We will also explore ways to keep students' identities safe and why it is necessary to protect one's identity.

Mathematics

Visual representations are mathematical tools often used to organize data (MP.5). A Venn diagram is a diagram that shows relationships among sets of data. In the case discussed in this chapter, three circles are used to represent each of three sets. They are overlapped in a way that shows relationships or commonalities between each set. Students will classify data as either being personal, shared with parents, shared with friends, or in the overlap between groups (K.MD.B.3; 1.MD.C.4). Though the discussion will not be based on numbers, it is still important for students to be able to justify their reasoning for classification and critique the reasoning of others (MP.3).

Financial Literacy

As your students grow up in a world of social media, which can be a place for publicly sharing an abundant amount of information about a person, it is important to talk with them about keeping private information safe. Students need to be part of discussions regarding why it is important not to share information with strangers and what they can do to help protect their privacy. Part of being a financially literate person is knowing what information can be shared and with whom. In the activity in this chapter, students are encouraged to explore different types of personal information and decide which would be appropriate to share with family and friends. Because of the age level of K–grade 5 students, we do not recommend the public sharing of any information on social media or other technology-related outlets.

Student Knowledge and Experience

Most students will have had some sort of previous experience regarding social media. Parents might have Twitter or Facebook accounts. Older siblings or family members might post to Instagram or Snapchat, and some students will even have their own devices at home. Though students may be aware that social media exist, they might not be as conscious of how easily private information can go public.

Lesson Plan

Learning Targets

- Organize and interpret data to answer questions.
- Explain how to keep private information private.

Resources and Tools

- **Activity Sheet 15.1:** Organizing Data (one for each student)
- **Resource 15.1:** Venn Diagram

Preparation Prior to Lesson:

Part 1: Create a larger version or project Resource 15.1 onto the board. Make enough copies of Activity Sheet 15.1 for each student.

Part 2: None.

Children's Literature Option

A Smart Girl's Guide: Digital World: How to Connect, Share, Play, and Keep Yourself Safe by Carrie Anton (2017). See the Literature for Financial Literacy list on p. 203 for additional ideas.

Key Language

personal information: Information that is about oneself.

private: Not meant to be shared with another individual.

Venn diagram: Diagram that shows relationships among sets of data.

Money Talks

Explain to students that **personal information** is information that is about oneself. The information is unique to each individual and can be used to differentiate us from others. Ask students to brainstorm generic personal information categories that can be used as data later in the activity. Answers might include: birthday, Social Security number, street address, telephone number, email address, and student ID. If students do not come up with it on their own, add bank account numbers to the list.

Sales Pitch

Which of this personal information should be considered **private**? Talk with students about private information being that which we don't want shared with other individuals. This might include secrets between friends or other personal information mentioned previously. Ask students, "Why might we want to keep some information private?"

Explain to students that they are going to use a Venn diagram to organize their information into categories, including self, parents, and friends. A **Venn diagram** is a diagram that shows relationships among sets of data and can lead to great discussion and justification explanations.

The Fine Print

Part 1: Venn Diagram

Create a larger version or project **Resource 15.1: Venn Diagram** onto the board. Talk with students about what each circle in the diagram represents. Explain the structure of a Venn diagram and that data that is placed in overlapping positions means it falls into both categories of circles.

Pass out **Activity Sheet 15.1: Organizing Data** so that each student has the opportunity to manipulate the data as the activity progresses. Explain to the students that the words on the left side of the page are going to be put into three categories: 1. Kept to Self, 2. Shared with Parents, 3. Shared with Friends. If the student believes the data to overlap categories, then they would place the word in the correct position on the diagram. Go over each word with the students to ensure that they understand its meaning.

Give students the opportunity to cut out the words and place them where they see fit. After an ample amount of time, have students share their classification with a partner to look for consistencies. If students disagree on the placement of a word, encourage them to justify their reasoning for their answer.

Standards for Mathematical Practice: Students need to be able to verbally explain their reasoning and have a productive discussion when others disagree with their answer (MP.3).

Part 2: Class View

After students have had time to discuss with a partner, bring the class together to create a class view of the Venn diagram. Encourage students to justify why they chose to locate the item in the specific position. Ask the following questions:

- Why do you believe the item belongs in this category?
- [If it is only one category] Why does it not belong in the others?
- How do I show if an item belongs in all three categories?
- When would this ever occur?

Give students time to discuss the various scenarios and come up with a consensus for categorizing each item. Remind students that even when we tell someone something in confidence, it might not always stay private. Explain to students that it is important to protect your identity and to never give out any information on the Internet without consent from an adult. This includes photos as well as written information!

Closing the Deal

Ask students if they thought using a Venn diagram was an easy way to organize information. See if they can apply this knowledge to another set of data using this tool. (For instance, weather. Sometimes it is just sunny or just raining, but other times it is both.)

Getting Your Money's Worth

The discussions in this activity can be modified based on the grade level and maturity of the students. Regardless of their ages, it is important to talk with students about keeping their information safe and when it is appropriate and not appropriate to share information with others.

References and Resources

Anton, Carrie. *A Smart Girl's Guide: Digital World: How to Connect, Share, Play, and Keep Yourself Safe*. Middleton, Wisc.: American Girl, 2017.

Jump$tart Coalition for Personal Financial Literacy. *National Standards in K–12 Personal Finance Education.* Fourth Edition. Second Printing. Washington, D.C.: Jump$tart Coalition for Personal Financial Literacy, 2017.

National Governors Association Center for Best Practices and Council of Chief State School Officers (NGA Center and CCSSO). *Common Core State Standards for Mathematics.* Washington, D.C.: NGA Center and CCSSO, 2010.

Activity Sheet 15.1

Organizing Data

Birthday	Phone number
Home address	Email address
Student ID	A secret between friends
Bank account information	Social Security number
Parent or guardian's phone number	Parent or guardian's address
Favorite color	Age

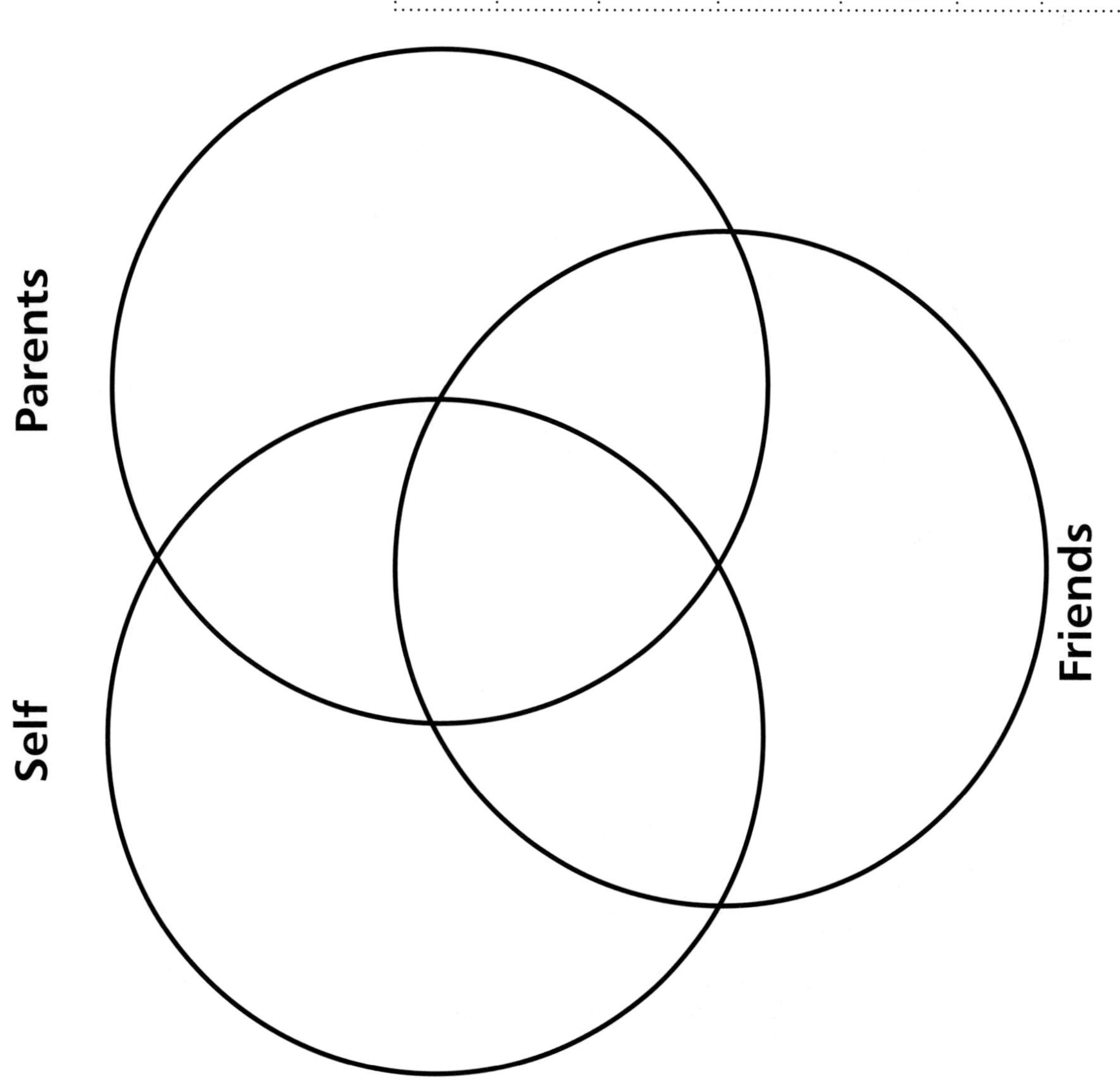

Resource 15.1

Venn Diagram

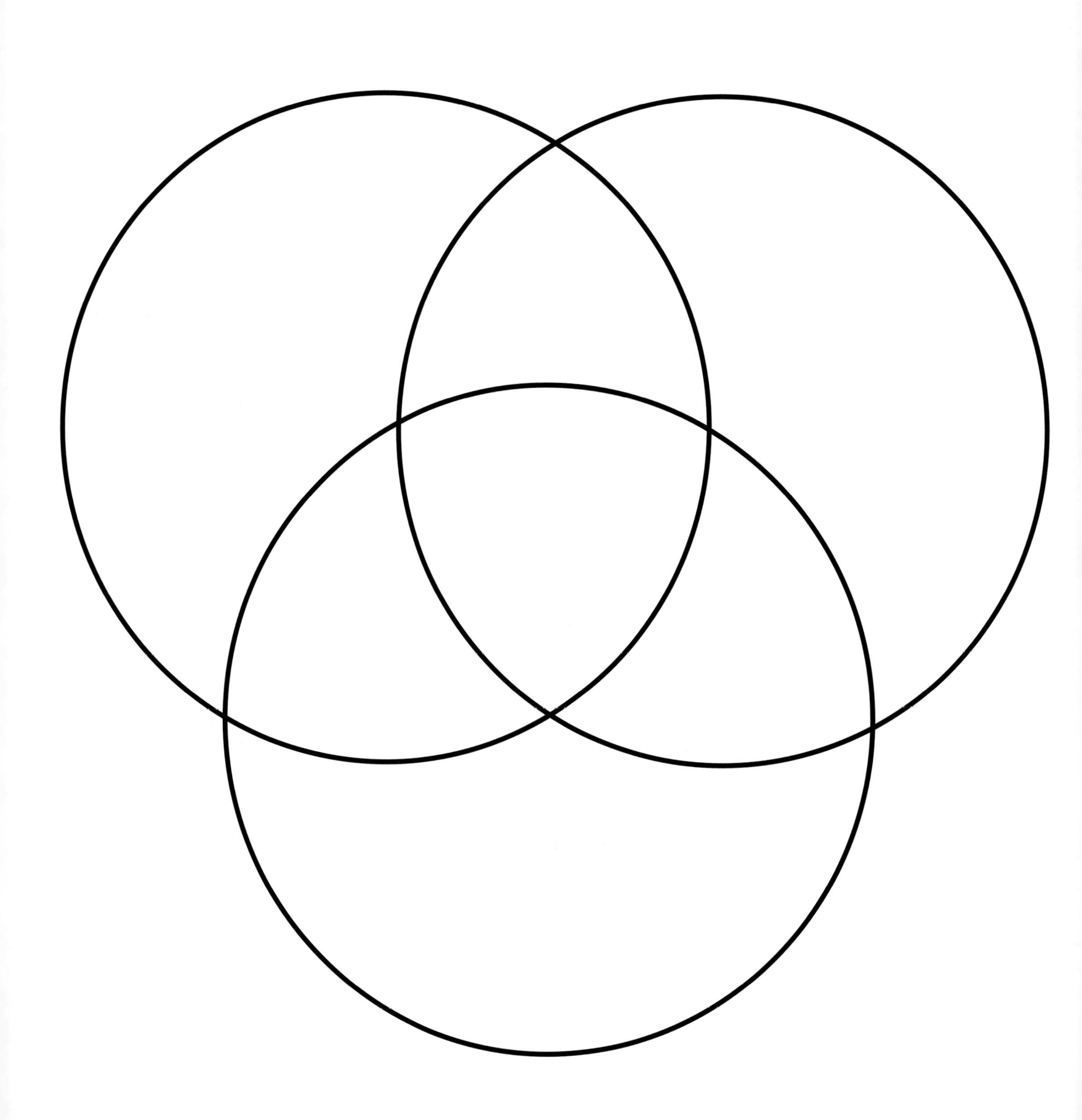

Chapter

How Does It Stack Up?

The Bottom Line

Money can be used as a mathematical tool to explore measurement.

Students use money as a nonstandard unit of measurement to explore height, width, and length. They will also count coins to find the total value of the measurement.

Standards Met in This Chapter

CCSSM Standards and Practices

K.MD.A.1: Describe measurable attributes of objects, such as length or weight. Describe several measurable attributes of a single object.
1.MD.A.2: Express the length of an object as a whole number of length units, by laying multiple copies of a shorter object (the length unit) end to end; understand that the length measurement of an object is the number of same-size length units that span it with no gaps or overlaps. Limit to contexts where the object being measured is spanned by a whole number of length units with no gaps or overlaps.
2.MD.A.2: Measure the length of an object twice, using length units of different lengths for the two measurements; describe how the two measurements relate to the size of the unit chosen.
MP.4: Model with mathematics
MP.5: Use appropriate tools strategically
MP.6: Attend to precision

(National Governors Association Center for Best Practices and Council of Chief State School Officers [NGA Center and CCSSO] 2010)

Jump$tart Standards

Spending and Savings: Apply strategies to monitor income and expenses, plan for spending and save for future goals.
Standard 3. Describe how to use different payment methods.

Knowledge Statements:
Overall:

- Paper money and coins have different values.

Standard 3:

- Count items up to 10.

(Jump$tart Coalition for Personal Financial Literacy 2017)

Future Value

It is important for students to be able to conduct accurate measurements using standard and nonstandard units. This activity begins with nonstandard units, but it can be modified when students are ready to measure in customary and metric units. It also provides the foundation for future money skills—e.g., rolling coins to exchange for bills—as a method for collecting loose change when saving for a prospective purchase.

The following sections describe the mathematical and financial literacy concepts that this lesson will encompass, as well as the knowledge and experiences that elementary students may bring to the lesson and take from it on its completion.

Did you know?
According to the Guinness World Records website (n.d.), the Burnt Hickory Youth Ministry in Marietta, Georgia, was able to lay down a mile of coins in a time of 2 hours, 16 minutes, and 9 seconds. In doing so, they were also able to fundraise $16,000 for humanitarian aid for Honduras!

Mathematics

In this chapter, students are able to explore the measureable attributes of various coins, which can lead to a rich discussion incorporating mathematical vocabulary, such as height, length, and width (K.MD. A.1). Students will also have the opportunity to practice fine motor skills through the act of stacking, both vertically and horizontally (in a row). When creating rows, standard 1.MD.A.2 addresses the importance of placing items in a way that does not lead to gaps between objects. This skill, along with the concept of measuring objects using appropriate tools, will set the foundation for good practice when introducing the basics of measurement (1.MD.A.2).

Using both nonstandard and standard units of measurement affords the students the ability to construct their knowledge of units and to see how variances in measurement relate to differences in unit size (2.MD.A.2). Students will be encouraged to measure more than once to explore unit size and also to attend to precision (MP.6). Through measuring everyday objects in multiple ways, students have the luxury of working with money, while practicing mathematical skills related to measurement (MP.4; MP.5).

Financial Literacy

In these activities, students are engaged in skills that use money as a nonstandard unit. This provides them with the reminder that coins have different attributes and values as well as reinforcing the skills of identification and worth. Students are also encouraged to think about saving coins and how this habit can accumulate money over time. They are taught how to organize their coins in bank-issued coin rolls, which is one way to keep track of their money so it does not get lost.

Student Knowledge and Experience

Kindergarten through grade 2 students might be new to the concept of saving money, but connections can be made to the real-life experience of other activities, such as saving tickets at Chuck E. Cheese or a carnival so that they can trade for the larger prize, or saving stickers on an achievement chart. Though the focus of this chapter is on measurement, the more experience students have with working with money, the more likely they are to recognize and be able to name the value of the coins.

Lesson Plan

Learning Targets

- Use pennies as a nonstandard unit of measurement.
- Accurately measure the height, length, and width of various objects.
- Identify measureable attributes of coins.
- Count a set of like coins and identify their total value.

Resources and Tools

- **Activity Sheet 16.1:** Collecting the Data (one for each student or pair of students)
- **Resource 16.1:** Coin Roll Chart (one projected for the class, or one copy for each student or group of students)
- **Resource 16.2:** Footprint
- Coins (at least 30 pennies per group)

Preparation Prior to Lesson

Money Talks: Display Resource 16.1 for the students or prepare enough copies for each student or groups of students.

Part 1: Display Resource 16.2 for the students or prepare enough copies for each student. Prepare bags of pennies with at least 30 per bag.

Part 2: Prepare enough copies of Activity Sheet 16.1 for each student.

Children's Literature Option

See the Literature for Financial Literacy list on p. 203 for relevant children's books.

Key Language

horizontal: Relating to the horizon; level or flat.

length, width, height, long, short, wide, tall, top, bottom, side: Words used in measurement .

measurement: A size, length, or amount of something; the act of measuring.

nonstandard unit of measurement: An informal unit of measurement made up of objects such as toothpicks, paper clips, or cubes.

standard unit of measurement: A unit of measurement defined by convention or law; e.g., customary or metric.

vertical: Raised or directed upward.

Money Talks

Frequently when we deposit an abundant amount of coins into the bank, we use coin rolls to organize them. (Show an example of a coin roll. Many banks will provide them to customers for free.) Ask students the following questions:

- Do you think all of these coin rolls are the same size?
- If not, which one do you think is larger or smaller and why? (Discuss various coin attributes.)
- Estimate how many coins each roll will hold.

The number of coins in standard coin rolls in the United States, as stated by Headley (2017), is located on **Resource 16.1: Coin Roll Chart.** Display Resource 16.1 for the students (without the answer key) and see if they can work together to fill in the missing parts.

Sales Pitch

Begin with a discussion on **measurement** and what it means to measure something. Talk about tools we use for measuring objects and the difference between **standard** and **nonstandard units of measurement**. Explain to the students that they are going to trace their foot on a piece of paper and estimate, or guess, how many pennies long they think their foot will measure (**Resource 16.2: Footprint** can be used as an example). Discuss assumptions that need to be clarified before the task, such as measuring from big toe to heel, measuring in a straight line, making sure there are no gaps between coins, and whether to round up or down if part of a penny is included.

Make a class chart of estimates and talk about why they are different. Through the class discussion, guide students to the realization that feet are different lengths and some will measure longer than others. The constant in this beginning exercise is that the size of the penny does not change. Depending on the age level of your students, you can estimate length with nickels, dimes, and quarters as well. See if students can make the connection between the larger-sized coin having a smaller quantity when measuring foot length than the smaller coin. Next, add to the class chart the actual length of each foot and how much that would be if they used skip-counting or added their coins by value.

The Fine Print

Part 1: Measurement Vocabulary

Place a tissue box or another classroom item in front of the students. Have them brainstorm ways to measure the box using their pennies. Record words on the board such as **long, short, tall, wide, high, top, bottom**, and **side** as students offer up ideas. If the words **length, width**, or **height** are not mentioned, place these three words on the board. Ask students which of their brainstormed words might fit into each of these three categories.

Have students demonstrate how they would use the pennies to measure the box. Something they will need to consider is whether they will be using a **vertical** stack or a **horizontal** stack. A horizontal stack (or row) is one like that used to measure the foot. The reason we call it a horizontal "stack" is because sometimes people stack carts or other items that do not sit on top of each other, but rather sit beside or nestled into each other. An example of this could be a set of books tightly placed between two bookends. A vertical stack would be one in which the pennies are stacked on top of each other, and the measurement depends on the thickness of the coin, not the diameter.

Standards for Mathematical Practice: Students develop MP.8 as they look for patterns for multiplying or dividing.

Model a horizontal stack measurement by setting pennies in a row on the top of a tissue box using sticky tack on the back of each coin. Adhere them to the box in a straight line, making sure that the coins are touching but do not overlap. Record your data on a chart as an example for the students to use in the next part of the activity. Instruct students to always label their units. In this instance, the unit would be pennies. You can also repeat this activity using vertical stacks.

Part 2: Collecting the Data

Depending on the grade level you are teaching, this activity can be adapted to include pictures or words. **Activity Sheet 16.1: Collecting Data** can be used to collect the data; students could also make their own data-collection sheet using pictures, numbers, and words. Instruct students to find three to five items in the room that they can measure with pennies. Have them collect these data with a partner, encouraging discussion around measurement, mathematical practices, and discoveries. Have students record their data, making sure to label their measurement as one of length, width, or height, and the value of their measurement in money.

Part 3: Extending the Data

Now that students have had an opportunity to explore their own items with partners, come together as a whole group and discuss the data. Ask students, based on their noticings, to estimate how many pennies they think it would take to measure the length of the classroom (or something significantly smaller depending on the grade level).

Give the students time to discuss with small groups how they might accomplish such a large task. Remind the students that they do not have enough pennies to lay them all out, so they need to find a strategy to measure a smaller portion and use the measurements to apply to the larger length.

As students are working in groups, ask them to explain their thinking to others and defend their choices for strategic actions. This activity will most likely extend over several days, so don't rush the students' possibility for creativity!

Closing the Deal

When the groups have had time to come to agreement on the length of the classroom, record each group's data in a class chart. Discuss with students why the measurements might differ. Where is the room for error in the measurements? Ideas might include that the line wasn't straight, there were gaps or overlaps in the pennies, the smaller measurement was inaccurate, or the calculations for the larger measurement were incorrect. Explain the need to measure more than once to ensure accuracy in our data. Choose one or more strategies to measure the actual length of the room, and compare the results to the class chart.

Getting Your Money's Worth

This activity promotes problem solving through exploration and discovery. Students can be further challenged with questions such as the following:

- Which do you believe would be longer: a row of four quarters or a row of eight dimes?
- Which do you believe would add up to more money; an inch of vertically stacked nickels or a centimeter of vertically stacked dimes?

Students can make connections to the concept of saving and how we roll coins when we want to organize them for exchange or deposit. You could share an example of a time that you saved coins in a jar or another safe place until you had enough money to make a desired purchase. Discuss with students the benefit of keeping their money in a safe place and waiting until they collected enough money for something larger rather than spending it as soon as they get it. Refer back to bank-issued coin rolls and ask them why rolling coins could be helpful. Ideas might include that it is easier to keep track of, easier to count, or less likely to spend.

References and Resources

Guinness World Records [website]. "Fastest Mile of Coins" (n.d.). http://www.guinnessworldrecords.com/world-records/fastest-mile-of-pennies.

Headley, Susan. "How Many Coins Are in a Regular Roll of Coins?" The Spruce Crafts [website] (2017). https://www.thesprucecrafts.com/number-of-coins-in-roll-768862.

Jump$tart Coalition for Personal Financial Literacy. *National Standards in K–12 Personal Finance Education*. Fourth Edition. Second Printing. Washington, D.C.: Jump$tart Coalition for Personal Financial Literacy, 2017.

National Governors Association Center for Best Practices and Council of Chief State School Officers (NGA Center and CCSSO). *Common Core State Standards for Mathematics*. Washington, D.C.: NGA Center and CCSSO, 2010.

Activity Sheet 16.1

Collecting the Data

Item	Unit of Measurement	Height, Width, or Length?	Actual Measurement	Total Value Amount
Example: *tissue box*	*pennies*	*length*	*12 pennies*	*12¢*
1.				
2.				
3.				
4.				
5.				

Notes:

Item	Unit of Measurement	Height, Width, or Length?	Actual Measurement	Total Value Amount
Example: *tissue box*	*pennies*	*length*	*12 pennies*	*12¢*
1.				
2.				
3.				
4.				
5.				

Notes:

Resource 16.1

Coin Roll Chart

Instructions: Fill in the missing information in the chart.

Denomination	Number of Coins	Value
	50	$0.50
Nickel		$2.00
Dime	50	
	40	$10.00
Half-Dollar	20	
One Dollar		$25.00

Answer Key:

Denomination	Number of Coins	Value
Penny	50	$0.50
Nickel	40	$2.00
Dime	50	$5.00
Quarter	40	$10.00
Half-Dollar	20	$10.00
One Dollar	25	$25.00

Resource 16.2

Footprint

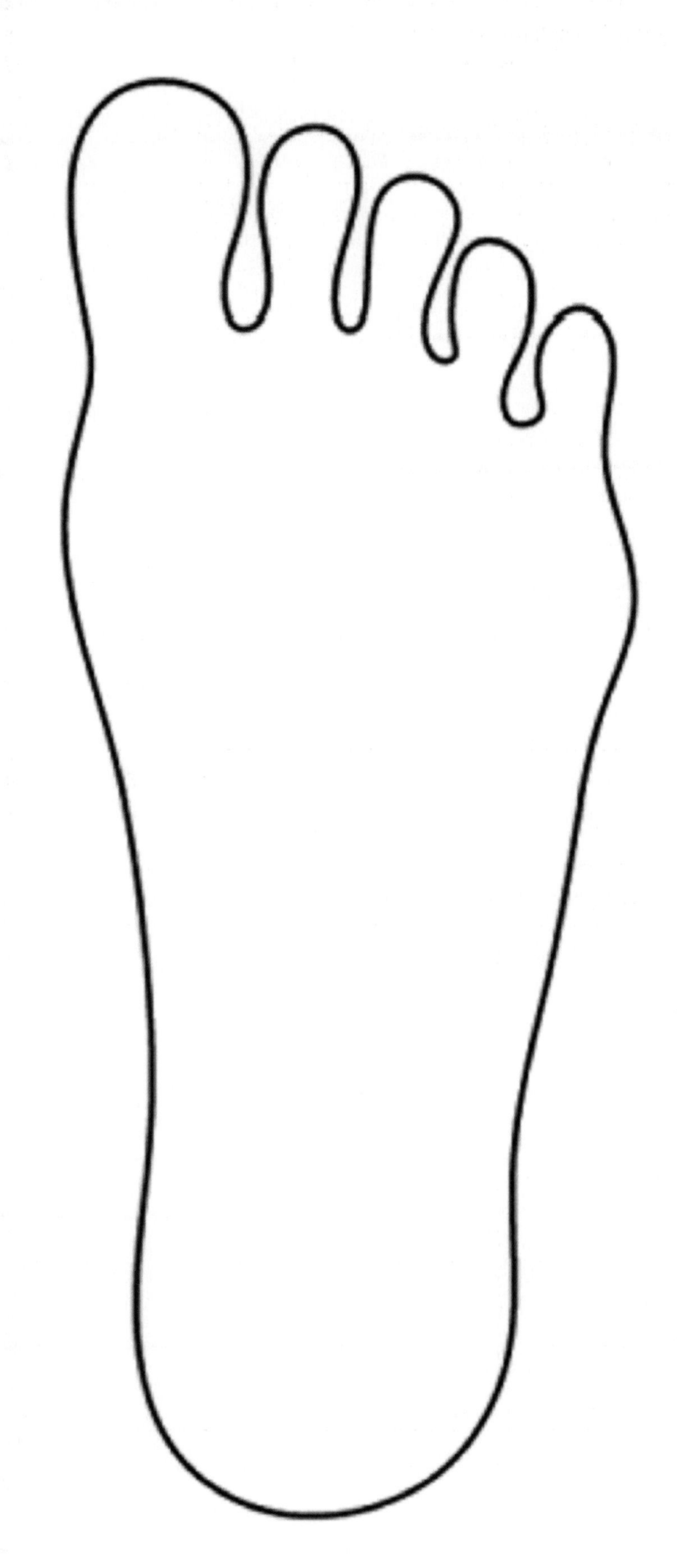

Chapter

Show Me the Money!

The Bottom Line

Students can use data to consider future job wages and career opportunities.

Students explore jobs that they are able to do (or soon will be able to do). They create line plots to examine hourly rates of various job opportunities.

Standards Met in This Chapter

CCSSM Standards and Practices

3.MD.B.3: Draw a scaled picture graph and a scaled bar graph to represent a data set with several categories. Solve one- and two-step "how many more" and "how many less" problems using information presented in scaled bar graphs. *For example, draw a bar graph in which each square in the bar graph might represent 5 pets.*

4.MD.B.4: Make a line plot to display a data set of measurements in fractions of a unit (1/2, 1/4, 1/8). Solve problems involving addition and subtraction of fractions by using information presented in line plots. *For example, from a line plot find and interpret the difference in length between the longest and shortest specimens in an insect collection.*

5.MD.B.2: Make a line plot to display a data set of measurements in fractions of a unit (1/2, 1/4, 1/8). Use operations on fractions for this grade to solve problems involving information presented in line plots. *For example, given different measurements of liquid in identical beakers, find the amount of liquid each beaker would contain if the total amount in all the beakers were redistributed equally.*

MP.3: Construct viable arguments and critique the reasoning of others.

MP.4: Model with mathematics.

(National Governors Association Center for Best Practices and Council of Chief State School Officers [NGA Center and CCSSO] 2010)

Jump$tart Standards

Employment and Income: Use a career plan to develop personal income potential.
Standard 1. Explore job and career options.
Standard 2. Compare sources of personal income and compensation.

Knowledge Statements:
Overall:

- People can earn income in the form of wages, salaries, and tips.

Standard 1:

- Give examples of careers related to personal interest.
- Explain the difference between a career and a job.

Standard 2:

- Identify jobs children can do to earn money.
- Explain the difference between a wage and a salary.

(Jump$tart Coalition for Personal Financial Literacy 2017)

Future Value

In this chapter, students will explore various careers and jobs and how the income from these differs. Through it is not necessary to choose a career while still a young student, discussing the options and characteristics of various jobs communicates that there is a wide range of opportunities. It is important for students to understand that careers vary greatly in terms of skills needed, education required, and potential earnings. Planting these seeds early helps students pay attention to the many adults they encounter in their daily lives, thinking about what skills and education these people have to hold such positions, as well as what their related income might be.

The following sections describe the mathematical and financial literacy concepts that this lesson will encompass, as well as the knowledge and experiences that elementary students may bring to the lesson and take from it upon its completion.

Mathematics

Students will explore graphs and the ways that information can be displayed and interpreted. To be statistically literate, students need to know which type of statistical display best allows them to analyze their data. Students need to know how to create categorical displays (e.g., bar graphs and picture graphs) and numerical displays (e.g., line plots), recognizing when each type of graph is a good fit for the data they have collected.

Categorical data, which is explored in part 1 of the lesson, is the focus of grade 3 data standards, but it is also important to grade 4 and 5 students so that they can contrast categorical and numeric data (3.MD.B.3). In part 1, students gather categorical data about career choices and determine which type of graph they want to make and which scale to use to represent their data (MP.3 and MP.4)

In part 2, students explore data on wages for jobs they can do and create line plots and look at the shape of the data (4.MD.B.4; 5.MD.B.2). A line plot can be skewed left (more data to the left), skewed right (more data to the right), symmetrical (general balance), or bimodal (two different clusters of data).

The focus of this investigation is on the process of working with statistics. As in chapter 14, this lesson incorporates the four-step process of doing statistics (see fig. 17.1). Students will create line plots to analyze numerical data (step 3) in order to reflect on and answer the statistical question (step 4) of how much money students make in various jobs (MP.4).

I. Formulate Questions

- Clarify the problem at hand.

II. Formulate one (or more) questions that can be answered with data.

- Collect data.
- Design a plan to collect appropriate data.
- Employ the plan to collect data.

III. Analyze Data

- Select appropriate graphical and numerical models.
- Use these methods to analyze the data.

IV. Interpret Results

- Interpret the analysis.
- Relate the interpretation to the original question.

Source: Franklin et al. 2005, p. 11.

Fig. 17.1. Statistical problem-solving framework

Financial Literacy

Two distinctions are made in this chapter: salary versus wages, and careers versus jobs. The Jump$tart Coalition for Personal Financial Literacy makes a distinction between a wage and a salary (Standard 2: Identify sources of personal income). While both are types of compensation, a wage is usually calculated on an hourly, daily, or piecework basis and paid on a schedule such as weekly, biweekly, or monthly. For example, a teen might get paid $10 an hour to mow a lawn and would be paid biweekly by the lawn owner. A salary is expressed as an annual sum and paid in prorated portions, which are usually weekly, biweekly, or monthly. Students likely understand salaries in terms of how much money a particular profession might make in a year (the focus of the first activity). Students will benefit from exploring wage options that are available to them as young people (these are considered occasional jobs). Students will explore various occupation opportunities to become more familiar with jobs and careers that interest them and the wages or salaries that accompany these positions.

Did you know?
In a Care.com 2017 Babysitter Survey, the average babysitting rate in 2016 was $13.97 per hour (up 26 percent from $11.11 per hour in 2010). Rates vary from city to city. The highest? San Jose, California at $16.68 per hour. The lowest? Youngstown, Ohio at $11.81 per hour.

Student Knowledge and Experience

In this lesson, it is important to select careers that are familiar to and of interest to students. Because they may only think of careers that are visible to them (e.g., teaching or a job their parents have), you might show a list of possible careers or provide time for students to explore career opportunities through the resources provided within the lesson. While students might see salary as a major determinant in choosing a career, they should also be exposed to other aspects, such as job description and accompanying responsibilities.

Did you know?
In 1938, the federal minimum wage in the United States was $0.25! (U.S. Department of Labor n.d.).

As students view the salaries that each career has, they have an opportunity to compare these salaries. This lesson therefore provides an opportunity to compare larger numbers in an authentic context. Students might notice the difference between salaries (difference comparison) or how many times greater one job's salary is than another (multiplicative comparison). Both types of reasoning are important and provide excellent preparation for the study of ratios and proportions in middle school. This lesson builds on students' prior knowledge of what it means to work and earn money.

Part 2 of the lesson assumes that students know the equivalencies of common decimals and fractions used with money (0.25 = 1/4; 0.50 = 1/2; 0.75 = 3/4).

Lesson Plan

Learning Targets

- Display categorical and numerical data in appropriate graphical displays.
- Explore the differences between careers and jobs, as well as between wages and salaries.
- Organize and interpret data to ask and answer questions related to job wages.

Resources and Tools

- **Activity Sheet 17.1:** Show Me the Money!
- 1-inch grid graph paper
- Markers

Preparation Prior to Lesson

Part 2: Graph paper (optional) for each student or pairs of students.

Part 2: Copy Activity Sheet 17.1 for each student or pair of students.

Children's Literature Option

Lemonade in Winter: A Book about Two Kids Counting Money by Emily Jenkins and G. Brian Karas. See the Literature for Financial Literacy list on p. 203 for additional ideas.

Key Language

career: An occupation that a person usually has for a significant time in their life and includes opportunities for advancement.

categorical data: Data that are topics or objects (not numbers)

data: pieces of information

job: Work for which a person is paid.

numerical data: Data that are numbers (e.g., prices, ages).

salary: A fixed regular payment, typically biweekly or monthly, earned by an employee.

wage: A fixed regular payment, typically daily or weekly, earned by an employee.

Money Talks

Ask students if they have thought about what they might like to do as a **career** when they grow up. Talk with them about a career being something that usually lasts longer than a **job** and one that likely requires advanced training or schooling. Call on several students to share and list various careers on the board. Pick a few examples and ask the students what they know about that particular career. Ask them questions such as the following:

- Do you have to go to college to have an occupation in this field?
- What are some of the responsibilities of someone that holds this position?
- Would you like to do this line of work for a short or long period of time and why?

Discussing students' interests will help them start thinking about careers that match their strengths and ways they might improve to become better learners in order to succeed in such a career.

Sales Pitch

Ask students to think about a job they have done in which they made money. If they have not had such an opportunity, ask to think of one way they could earn money now, at their current age. Then have them think of something that they might want to do as an adult. In groups of three to four students, have them share each of their answers for earning money now and for earning money in the future. Ask each group to come up with at least one way that money they earn now is like how they might earn money in the future and at least one way the money they earn now is different from what they want to do in the future. Record ideas on a board or a large piece of paper, as shown in fig. 17.2.

Money Earning Now and in the Future	
How are they alike?	How are they different?

Fig. 17.2. Ideas for earning money now and in the future

The Fine Print

Part 1: Career Explorations

Ask students to match these careers with their annual salaries (Bureau of Labor Statistics 2016):

1.	Truck drivers	a. $20,580
2.	Computer programmer	b. $138,190
3.	Counselor	c. $49,740
4.	Elementary teacher	d. $85,180
5.	Orthodontist	e. $96,440
6.	Engineer	f. $38,900
7.	Fast food cook	g. $228,780
8.	Lawyer	h. $57,470

Correct answers: 1-f, 2-d, 3-c, 4-h, 5-g, 6-e, 7-a, 8-b

Explain that these figures are **salaries**. They are not hourly rates or **wages**. Ask students to pair-share to consider how a salary is different from a wage. Discuss the pros and cons of being paid a salary (flat rate) or wage (an hourly rate). Next compare careers and jobs.

Clarify that students will not be making money like this until they finish school and start a career. Refer to the differences that students listed in comparing a job now to one in the future. Explain that a career is different than a job. A career is something that they are trained to do and that they will likely do for a long time. Jobs might also require training and last a long time, but jobs can also be very short-term.

When selecting a career, it is also important to consider what you like to do and what you are good at doing. If time allows and technology is available, have students investigate career opportunities through a website. Two good options are the following:

- **Career Exploration** on the Bureau of Labor Statistics website (https://www.bls.gov/k12/content/students/careers/career-exploration.htm).
- **Salary Data & Career Research Center** on the PayScale website (https://www.payscale.com/research/US/Country=United_States/Salary)

Alternatively, you can post a list of career options. The purpose of this activity is to get students thinking about the various lines of work and the description of what this work entails. It is not to encourage students to pick work based solely on salary, though it is important for students to know that salaries are very different for different careers (and why that might be). Picking a career can be a big decision. There are many qualities of a career that one must consider, including salaries and benefits, level of difficulty, social interactions, autonomy, and physical environment (Smith 2018). To give students some perspective into each career, they will be doing research on various fields of their choice.

Standards for Mathematical Practice: Students demonstrate MP.4 as they determine how to represent the data in a graph.

Invite students to think of their top three favorite careers. This will provide a lot of categorical **data**. Have students write each idea on a sticky note and place the notes in the middle of their table or group of desks. Ask the following questions:

- What do you notice about our data?
- What kind of data is this—**categorical** or **numerical**?
- What options do we have for organizing categorical data?

Students may notice that they will need to combine some ideas into categories. Discuss possible categories (e.g., if there are a number of careers within the medical profession, a category could be Medical Careers).

Assign each group of two to four students the task of making a pictograph or bar graph that is scaled. Explain that a scaled graph means that each picture represents a set number of data points. If they decide that one box (bar graph) or picture (pictograph) represents four pieces of data, then they only create one box or picture for four data points. If there are ten responses, that is two full boxes or pictures, plus half of one box or picture.

When they are completed, ask students to share the graphs they have made. Compare the different graphs to examine how they captured the data. Ask the following questions:

- How do we figure out how many people picked [a medical profession] on this group's graph? On this group's graph?
- Why might one group have more pictures (longer bars) than another group?
- Why might a bar graph be a good choice? Why might a pictograph be a good choice?
- What do you notice about the career choices from our class?

Summarize by asking students to share one thing they learned about careers and one thing they wonder about a career that they like. If time allows, have students share these thoughts.

Standards for Mathematical Practice: Students demonstrate MP.3 as they justify which graph they believe is a better fit and as they determine what scale makes sense for the data they have.

Part 2: Show Me The Money—Jobs

Ask students to revisit the list of jobs they have done to earn money. Explain that in this activity they will use data to understand the wages a young person might make from a job. Ask, "What do you think the going hourly wage is for a babysitter?" Share that the average babysitting rate in 2016 was $13.97 per hour. (You may also want to look up the average rate in your area and use that data.)

Explain that they are going to use the four-step process of a statistics investigation to answer the statistics question. First, a statistical question. Our question is, "How much money do preteens/teens make from their jobs?" Step 2 is collecting data. Engage students in a discussion of how this data could be collected in a survey; in particular, focus on how the question should be asked in order to obtain valid and useful data. Allow students time to discuss various ideas (which can be implemented and used instead of the data provided in this lesson's activity sheet).

Distribute **Activity Sheet 17.1: Show Me the Money!** Ask students to review the data. Explain that they are going to be analyzing the data (step 3) by using **line plots**. First have them create the line plots for questions 1–5. Then have them to come together, and ask, "What is meant by 'shape of data'?"

Standards for Mathematical Practice: Students are able to model with mathematics by using graphs to analyze information and draw conclusions based on the data (MP.4).

Explain that the shape of data gives an at-a-glance summary of the data. Share the different terms statisticians use to describe shape: *left-skewed, right-skewed, symmetric*, and *bimodal*. Have students complete parts b and c for questions 1–5 on their activity sheets (step 4).

As students work, and after they have finished, ask the following questions:

- What can you tell us about students' wages from your graphs?
- What is the shape of data for [walking dogs]? How does this compare to the shape of data for [babysitting]?
- Why might there be different wages for these jobs?
- Why might babysitting be a higher paying job than walking dogs?
- How might you use these data to try to get a higher wage for a particular job?
- What factors influence the wage you might make for any one of these jobs?

Closing the Deal

Ask students to pair with another student, assigning the labels of student A and student B. Ask student A to share what he or she learned about careers, and ask student B to share what he or she learned about jobs. Repeat the exchange for salaries and wages.

Finally, ask students to reflect on what they learned from their data on jobs. From this reflection, have them prepare their article for the school newsletter (question 6 on Activity Sheet 17.1).

Getting Your Money's Worth

To extend and connect learning, visit the Bureau of Labor Statistics website for K–12 resources (https://www.bls.gov/k12/home.htm). There are posters, games, puzzles, activities, and lessons for teachers to use in their classrooms; and students can explore career opportunities.

References and Resources

Bureau of Labor Statistics (U.S. Department of Labor). "Career Exploration" [web page]. https://www.bls.gov/k12/content/students/careers/career-exploration.htm

Care.com [website]. "Care.com Releases 2017 Babysitter Survey Results" (2017). https://www.care.com/press-release-carecom-releases-2017-babysitter-survey-results-p1186-q88331797.html.

Franklin, Christine, Gary Kader, Denise Mewborn, Jerry Moreno, Roxy Peck, Mike Perry, and Richard Scheaffer. *Guidelines for Assessment and Instruction in Statistics Education (GAISE) Report: A Pre-K–12 Curriculum Framework*. Alexandria, Va: American Statistical Association, 2005.

International Labour Organization. (n.d.). "What Is a Minimum Wage." http://www.ilo.org/global/topics/wages/minimum-wages/definition/WCMS_439072/lang--en/index.htm

Jenkins, Emily, and G. Brian Karas. *Lemonade in Winter: A Book about Two Kids Counting Money*. New York: Random House, 2012.

Jump$tart Coalition for Personal Financial Literacy. *National Standards in K–12 Personal Finance Education. Fourth Edition*. Second Printing. Washington, D.C.: Jump$tart Coalition for Personal Financial Literacy, 2017.

National Governors Association Center for Best Practices and Council of Chief State School Officers (NGA Center and CCSSO). *Common Core State Standards for Mathematics*. Washington, D.C.: NGA Center and CCSSO, 2010.

PayScale. "Salary Data and Career Research Center." https://www.payscale.com/research/US/Country=United_States/Salary.

Smith, Jane. "5 Most Important Aspects of a Job." Work - Chron.com (2018). http://work.chron.com/5-important-aspects-job-6301.html.

United States Department of Labor. "History of Federal Minimum Wage Rates under the Fair Labor Standards Act, 1938–2009" (n.d.). https://www.dol.gov/whd/minwage/chart.htm.

Activity Sheet 17.1

Show Me the Money!

Thirty students were surveyed about the jobs they have. The table shows their responses to the questions "What job do you have to make money?" and "How much money do you earn in this job?"

	Job	Hourly Wage		Job	Hourly Wage
1.	Babysitting	$9.25	16.	Babysitting	$7.00
2.	Lawn mowing	$9.25	17.	Washing cars	$5.50
3.	Lawn mowing	$8.00	18.	Lawn mowing	$9.00
4.	Dog walking	$7.75	19.	Babysitting	$7.25
5.	Babysitting	$11.50	20.	Babysitting	$8.00
6.	Washing cars	$6.50	21.	Babysitting	$5.50
7.	Washing cars	$5.50	22.	Lawn mowing	$6.75
8.	Dog walking	$5.50	23.	Babysitting	$8.50
9.	Dog walking	$5.00	24.	Babysitting	$10.00
10.	Babysitting	$11.50	25.	Washing cars	$5.25
11.	Babysitting	$11.00	26.	Lawn mowing	$10.00
12.	Washing cars	$6.00	27.	Dog walking	$4.50
13.	Lawn mowing	$9.25	28.	Lawn mowing	$6.75
14.	Washing cars	$5.50	29.	Dog walking	$5.00
15.	Dog walking	$6.00	30.	Lawn mowing	$6.75

To explore "How much money do middle school students earn in their jobs?" use line plots!

1a. Create a line plot for **dog walking**:

1b. What is the shape of the data?

1c. What does the shape of the data tell you about the hourly wage for dog walking?

Activity Sheet 17.1—*page 2*

Show Me the Money!

2a. Create a line plot for **washing cars**:

2b. What is the shape of the data?

2c. What does the shape of the data tell you about the hourly wage for washing cars?

3a. Create a line plot for **babysitting**:

3b. What is the shape of the data?

3c. What does the shape of the data tell you about the hourly wage for babysitting?

Activity Sheet 17.1—*page 3*

Show Me the Money!

4a. Create a line plot for **lawn mowing**:

4b. What is the shape of the data?

4c. What does the shape of the data tell you about the hourly wage for lawn mowing?

5. Create a line plot that shows wages for all the possible jobs:

5b. What is the shape of the data?

5c. What does the shape of the data tell you about the hourly wage students get paid?

Activity Sheet 17.1—*page 4*

Show Me the Money!

6. Your school newsletter wants you to write an article that will help students know about their job options. Use the data above to write 4 to 5 sentences that could appear in the school newsletter. Be creative!

Appendixes

Literature for Financial Literacy

The following list of children's books on financial topics is also available for download in a searchable and sortable Microsoft Excel file on NCTM's More4U website (http://www.nctm.org/more4u) by using the code on the title page of this book. The online version includes columns that tell which financial topics and activities (such as "save," "spend," "budget," or "coins") are touched upon within each book.

A Chair for My Mother by Vera Williams (Greenwillow 1982) (*K–Grade 2*)

A Dollar for Penny by Julie Glass (Random House 2000) (*K–Grade 2*)

A Dollar, a Penny, How Much and How Many? (*Math Is Categorical*) by Brian P. Cleary (Lerner 2012) (*K–Grade 3*)

A Princess Loves to Save by Carolyn Grossi (Little Flower Productions 2010) (*K–Grade 2*)

A Smart Girl's Guide to Money by Nancy Holyoke (American Girl 2014) (*Grade 5*)

Alex's Ten-Dollar Adventure by Wendy Bailey (Three Bean Press 2014) (*K–Grade 4*)

Alexander Who Used to Be Rich Last Sunday by Judith Viorst (Silver Burdett 1987) (*K–Grade 4*)

All About Money by Sharon Coan (Teacher Created Materials 2013) (*K–Grade 2*)

Ayo's Money Jar by Charlene Hill Fadirepo (CreateSpace 2016) (*K–Grade 1*)

Benny's Pennies by Pat Brisson (Dragonfly 1995) (*K–Grade 2*)

Brody Borrows Money by Lisa Bullard (Millbrook 2013) (*K–Grade 2*)

Coat of Many Colors by Dolly Parton (HarperCollins 1996) (*K–Grade 3*)

Counting with Common Cents: Penny's Journey by Deirdre McCarthy (CreateSpace 2014) (*K–Grade 1*)

Daphney Dollar's First Allowance by Sharon Lewis (CreateSpace 2015) (*K–Grade 3*)

Daphney Dollar's First Trip to the Store by Sharon Lewis (Fiscal Pink 2016) (*K–Grade 3*)

Deena's Lucky Penny by Barbara deRubertis (Kane Press 2006) (*K–Grade 1*)

DK Eyewitness Books: Money by Joe Cribb (DK Children 2016) (*Grades 3–5*)

Earning Money by Shelly Buchanan (Teacher Created Materials 2013) (*K–Grade 2*)

Finance for Kidz by Dr. Prakash L. Dheeriya (20 books on various topics) (Fintelligence Publishing) (*K–Grade 3*)

Financial Fitness for Life: Student Storybook by Donna K. Wright (National Council on Financial Education 2005) (*K–Grade 5*)

Follow the Money by Loreen Leedy (Holiday House 2003) (*K–Grade 3*)

Fun with Money by Dr. Carolyn Edwards (Self-published 2010) (*K–Grade 3*)

Growing Money: A Complete Investing Guide for Kids by Gail Karlitz (Price Stern Sloan 2010) (*Grades 4–5*)

Here's the Scoop by Neale S. Godfrey (Modern Curriculum Press 1995) (*K–Grade 3*)

How the Second Grade Got $8,205.50 to Visit the Statue of Liberty by Nathan Zimelman (Whitman 2017) (*Grades 2–5*)

How to Turn $100 into $1,000,000: Earn! Invest! Save! by James McKenna and Jeannine Glista (Workman 2016) (*Grades 4–5*)

I Can Be Anything I Dream by Dr. Carolyn Edwards (Self-published 2010) (*K–Grade 3*)

If You Made a Million by David M. Schwartz (HarperCollins 1994) (*Grades 1–5*)

It's a Habit, Sammy Rabbit by Sam X. Renick (It's a Habit 2001) (*K–Grade 3*)

Joe the Monkey Saves for a Goal by John Lanza (Snigglezoo 2011) (*K–Grade 2*)

Lemonade in Winter: A Book about Two Kids Counting Money by Emily Jenkins (Schwartz & Ward 2012) (*K–Grade 2*)

Lots and Lots of Coins by Margarette S. Reid (Dutton 2011) (*K–Grade 3*)

Making Money Grow by Kathleen E. Bradley (Teacher Created Materials 2013) (*Grades 4–5*)

Mighty Mommies and Their Amazing Jobs: A STEM Career Book for Kids by Donald Jacobsen (2017) (*K–Grade 2*)

Money Madness by David A. Adler (Holiday House 2009) (*K–Grade 3*)
Money-Making Sunny: Finance Whiz Kids by Darrah Brunstein (Finance Whiz Kids 2012) (*K–Grade 5*)
Money Mama & The Three Little Pigs by Lori Mackey (P4K Publishing 2004) (*K–Grade 5*)
Money Mammals for Your Family Kit by John Lanza (Snigglezoo 2018) (*K–Grade 3*)
Money Savvy Kid$ Club: Garage Sale Riches by Susan Beacham and Lynnette Khalfani Cox (Advantage World Press 2008) (*K–Grade 2*)
Money Savvy Kid$ Club: Penny Power by Susan Beacham and Lynnette Khalfani Cox (Advantage World Press 2009) (*K–Grade 2*)
Money Savvy Kid$ Club: Putting the "Do" in Donate by Susan Beacham and Lynnette Khalfani Cox (Advantage World Press 2008) (*K–Grade 2*)
Money, Money, Honey Bunny! by Marilyn Sadler (Random House 2006) (*K–Grade 1*)
Moonjar: K–5 Classroom Kit (Moonjar 2018) (*K–Grade 5*)
Ms. Pig E. Banks by Ajshay James (Self-published 2013) (*K–Grade 5*)
My Daddy, the Amazing Nurse!: A Rhyming Career Exploration Book for Children and Others by Donald Jacobsen (2017) (*K–Grade 2*)
Neale S. Godfrey's Ultimate Kids Money Book by Neale S. Godfrey (Simon & Schuster Children's Publishing 1998) (*Grades 3–5*)
Oh, the Places You'll Go! by Dr. Seuss (Random House 1990) (*Grades 1–5*)
Once Upon a Company . . . A True Story by Wendy Anderson Halperin (Orchard 1998) (*Grades 3–5*)
Once Upon a Dime: A Math Adventure by Nancy Allen (Charlesbridge 1999) (*K–Grade 1*)
One Cent, Two Cents, Old Cent, New Cent by Bonnie Worth (Random House 2008) (*K–Grade 2*)
Pennies by Mary Hill (Welcome Books 2005) (*K–Grade 1*)
Pigs Will Be Pigs: Fun with Math and Money by Amy Axelrod (Aladdin 1997) (*Grades 2–4*)
Rock, Brock, and the Savings Shock by Sheila Bair (Whitman 2017) (*Grades 3–5*)
Saving Salvatore by FUNancial Literacy LLC (*K–Grade 1*)
Shopping in the City: Subtraction Facts by Sara A. Johnson (*K–Grade 2*)
Sonja Meets Her Future Self by Women's Institute for Secure Retirement (WISER) (*K–Grade 4*)
Stock Market Pie: Grandma Helps Emily Make a Million by J. M. Seymour (DynaMinds 2002) (*Grades 3–5*)
The Coin Counting Book by Rozanne Lanczak Williams (Charlesbridge 2001) (*Grades 1–4*)
The Family Bank by Sergio Dinaro (AuthorHouse 2012) (*Grades 2–5*)
The Financial Fairy Tales: The Magic Magpie by Daniel Britton (CreateSpace 2016) (*Grades 1–5*)
The Go-Around Dollar by Barbara Adams and Joyce Zairns (Simon & Schuster 1992) (*K–Grade 3*)
The Lunch Line by Karen Berman Nagel (Cartwheel 1996) (*Grades 1–4*)
The Masloppy Family Goes to New York City by Catherine Twomey Fosnot (First Hand 2007) (*K–Grade 3*)
The Penny Pot by Stuart J. Murphy (HarperCollins 1998) (*Grades 1–3*)
The Power of the Penny by Elaina Redmond (Spencer 2009) (*K–Grade 5*)
The Saving Seed by Ashley Parks, CFP (Bb Madison Press 2011) (*K–Grade 5*)
The Story of Money by Betsy Maestro (HarperCollins 1995) (*Grades 3–5*)
Those Shoes by Maribeth Boelts (Candlewick 2009) (*K–Grade 5*)
Three Cups by Tony Townsley and Mark St. Germain (Thomas Nelson 2011) (*K–Grade 3*)
Using Decimals to Play Our Vacation Budget by Andrew Einspruch (Capstone 2010) (*Grades 3–5*)
What Are Budgets? by Andrew Einspruch (Teacher Created Materials 2008) (*Grades 3–5*)
When Times Are Tough by Yanitzia Canetti (BrickHouse 2010) (*K–Grade 3*)
Where Is My Money? by Twyla D. Prindle (PHP Inc. 2009) (*K–Grade 3*)
Why Is There Money? by Paul Nourigat (FarBeyond Publishing 2011) (*K–Grade 4*)
Why Money Was Invented by Neale S. Godfrey (Silver Press 1995) (*K–Grade 2*)
You Can't Buy a Dinosaur with a Dime by Harriet Ziefert (Blue Apple 2003) (*K–Grade 3*)

Financial Literacy Resources

Jump$tart's Coalition for Personal Financial Literacy

https://www.jumpstart.org

This organization published the National Standards in K–12 Personal Financial Education that are used for this book. Their site includes links to those complete standards as well as to a wealth of resources, publications, activities, and opportunities for educators and for students.

Practical Money Skills

http://www.practicalmoneyskills.com

This site contains student-friendly overviews of topics such as saving, credit cards, buying a car, and more. Students will also find games that mimic video games, but with financial literacy insights as the goal. The site also provides dozens of calculators for all things financial—from allowance comparison to developing a holiday budget, auto loan payments, and repaying student loans.

EconEdLink

http://www.econedlink.org

This site, provided by the Council for Economic Education, has a lot of information to strengthen your own understanding of personal finance, as well as that of your students. It includes lessons for all levels of K–12 and tools such as a compound interest calculator that can support students beyond the lessons provided on the site.

Junior Achievement

https://www.juniorachievement.org/web/programs/programs

This web page provides links to various programs for learning about personal finance, business, and careers, including one program—*JA BizTown,* a visit to a simulated town—aimed at students in the middle elementary grades.

Federal Reserve Education

http://www.federalreserveeducation.org

At this searchable web site, you can find many resources for students, educators, and adults. You can use filters to find all the material that is appropriate for elementary or middle school students.

Federal Deposit Insurance Corporation (FDIC)

http://www.fdic.gov/consumers/consumer/moneysmart/young.html

This site offers the Money Smart for Young People series, a group of financial education programs in four age bands that cover from pre–K to age twenty. The web page also includes a link to the Teacher Online Resource Center provided jointly by the FDIC and the Consumer Finance Protection Bureau (CFPB).

Consumer Financial Protection Bureau (CFPB)

http://www.consumerfinance.gov/consumer-tools/money as you grow/

This page on the CFPB website provides resources from the Money as You Grow program that is designed to help parents and caregivers give their children a sound financial education from age three through college and beyond.

Activity Sheets Answer Key

Many of the answers given by students on this book's activity sheets will vary, depending on each student's choices of purchases, jobs, and so on. The answers below are for questions without multiple correct answers, though some of these answers may be correctly expressed in more than one way (e.g., 31¢ or $0.31).

Activity Sheet 7.1

"Amount of money found" each day followed by "Total money saved":
Monday: 23¢; 23¢
Tuesday: 14¢; 37¢
Wednesday: 20¢; 57¢
Thursday: 13¢; 70¢
Friday: 7¢; 77¢
Saturday: 31¢; $1.08

Activity Sheet 8.1

Give and Take Challenges:

1. borrow
2. $42
3. $35

Activity Sheet 10.1

Tax rates in decimal form: 0.1; 0.15; 0.3; 0.35; 0.4

"Estimated Taxes" for each worker followed by "Estimated Net Pay":
Amalia: $860; $7740
Caroline: $15,000; $35,000
Johanna: $200,000; $300,000
Miguel: $27,000; $63,000
Jeremy: $5250; $29,750
Rebecca: $87,500; $162,500
Marie: $19,500; $45,500

Activity Sheet 11.1

6. $10
7. $1

Activity Sheet 11.2

1. add 200
2. 20; 30; 50
3. 16; 12; 4
4. times 8
6. $10
7. 10%

Activity Sheet 11.3

1. 0.5
2. 0.1
3. 0.01
4. $4.00
5. $40.00
6. $1040.00